Mystical Environmental Philosophy

How to Save Life on Earth

A catalogue record for this book is available from the British Library

ISBN: 978-1-907962-88-2

Publication Date: October 2021

Published by Cranmore Publications

Exeter, England

cranpubs@gmail.com

The life-forms of the Earth do cheer.

For, my Mysterious Prince is here!

Come out of your seclusion.

Free the human species from its delusion!

Use your unique insight, wisdom & learning.

Prevent this beautiful world from burning!

F. H.

The purpose of this book is to enable you to come to appreciate the force that propels the unfolding of the Universe. This unfolding encompasses everything, from the formation of planets, to the development of human culture. This force has inevitably brought forth the world that we live in today: a world dominated by technology, a world marked by the human transformation of the face of the Earth, a world which is in grave peril from the phenomenon of global warming.

The coming into being of this world was always on the cards. The coming into being of this world is not a freakish accident. This world, with its technology, its human transformation, and its global warming, is a part of the unfolding Universe, a part of our unfolding Solar System, as it progresses through the millennia to increasingly optimal states.

When one comes to know the force that propels the unfolding of the Universe, then one will know that attempting to deal with the phenomenon of global warming through fossil fuel emission reduction efforts is misplaced. Such efforts will achieve nothing good. Let us hope that they do not do too much harm, by distracting us from the actions that are needed to save life on Earth from the impending peril that it faces from the phenomenon of global warming.

If fossil fuel emission reduction efforts continue to be the mainstay of our response to global warming, the result will be significant global warming in the upcoming decades and centuries. This global warming will cause significant climate change, and the extreme flooding and heat events that this entails will, in turn, cause immense death and suffering that could have been avoided. If fossil fuel emission reduction efforts are the mainstay of our response to the phenomenon of global warming, then the ultimate result will be the obliteration of life on Earth.

The force that propels the unfolding of the Universe has already, inevitably, taken us to the place where fossil fuel emission reduction efforts are a hopeless distraction. Such efforts are analogous to attempting to extinguish a 2000 hectare blazing inferno by using the water that is in a child's water pistol. This is the tragic state that we are currently in; we have been deluded into believing that using the child's water pistol is a good idea. If we focus on fossil fuel emission reduction efforts as our strategy to avert global warming, and to avert death and suffering, then the result will be massive global warming and immense death and suffering. Immense death and suffering that could have been avoided if we retired the pistol and took a different approach.

We are cosmic puppets, who have temporarily been deluded into believing that we are fundamentally a destructive parasite, a destroyer of life. We have been temporarily deluded into believing that it is solely our actions that have put the planet in peril, and that we therefore need to act quickly to reign in our activities, to minimise our impact on the Earth, and thereby 'save the planet'. In reality, averting calamity for life on Earth requires us to simply embrace who we are, embrace what we are, to power forward, and to embrace our destiny.

Saving life on Earth requires us to deploy technology to actively control the Earth's atmospheric temperature. Saving life on Earth ultimately requires the deployment of solar-deflection technology at Lagrangian Point L1.

If we do not embrace out nature, our destiny, our cosmic purpose, then the result will be immense death and suffering in the coming decades and centuries.

It is not too late. We can change course and thereby achieve the best possible future for the human species and the rest of life on Earth. We can save the life that has arisen on the Earth.

Changing course seems to require the setting up of an organisation which has the purpose of helping us to fulfil our cosmic purpose. As our cosmic purpose is to deploy solar-deflection technology at Lagrangian Point L1, let us call this organisation:

The LP1 Foundation

If you would like to be part of The LP1 Foundation, and join this effort to save life on Earth, then it would be great to hear from you. You might have relevant expertise in an area such as aerospace engineering and/or you might be interested in providing funding for the foundation.

www.drcphilosophy.com

theuniversalpursuitofecstasy@gmail.com

Follow your passions and have a wonderful life

Dr C

How to Read this Book

The next part of this book – *Our Questions* – sets the scene for the rest of the book, so it would be a good idea to read this section first.

The rest of the book contains a number of sections which could be considered to be self-contained wholes. Yet, each section will be richly and fruitfully illuminated by all of its companions. You could read the rest of the book in the order in which it is presented, which is what I would suggest. However, there is no need to do this; if you find yourself being drawn to a particular section then you could start there.

Each section can be thought of as a piece of a jigsaw. My hope is that if you read the entire book, and thereby encounter all of the pieces, that everything will fall into place. My hope is that you will then have a transformation in your understanding, a transformation in your appreciation of the world around you. Until this moment of transformation, at the time when you have only encountered a few pieces of the jigsaw, things might not make much sense to you. This is an inevitable aspect of one's journey when one is on the path to having a transformation in one's understanding.

Our Questions

Life, what is it all about? Does the existence of the human species matter? In other words, is the human species cosmically significant, or is it just one species among many? If the human species is not just one species among many, why is it cosmically significant? What does it mean to be cosmically significant? How does cosmic significance relate to the nature of the Universe? How does the nature of the human species relate to the nature of the Universe? Why has the human species come to dominate the planet? Why has the human species acted so as to bring forth a plethora of pressing environmental issues, the most important of which is the phenomenon of human-induced global warming? These are our questions.

Our questions are outside the realm of science. Science does not encroach into questions of cosmic significance, or questions of the fundamental nature of that which it measures and predicts.

Our questions are also outside the realm of philosophy. Philosophy entails rational critique, but the conclusions that are reached, whilst intellectually plausible, and rigorously thought through, are either uninteresting trivialities or nothing but mere speculation. In the latter case, an alternative view can be just as intellectually plausible and robust.

Our questions are outside the realm of environmentalism. Environmentalists take the data provided by scientists and draw their own conclusions based on this data. This scientific data, the measurements and predictions, are wholly unimbued with anything relating to cosmic significance, or the fundamental nature of reality. Environmentalists can bring to the table their own personal philosophical worldview, but such a worldview is divorced from scientific data, whilst being grounded in philosophical speculation.

All of these things – science, philosophy, environmentalism – have value, they just don't enable our questions to be answered. The domain of our questions is the domain of mystical environmental philosophy. Mystical environmental philosophy is grounded in an immediate apprehension of the nature of the Universe. Once this mystical apprehension has occurred, our questions can gradually be answered, as this apprehension fuses with the environmental awareness that has arisen from science and the rational philosophical ability to synthesise and express.

This entire book is an expression of mystical environmental philosophy. You might love it, or be repulsed by it, or be wholly indifferent to it. I hope that you love reading it as much as I have loved writing it. However, all of these responses are fine, because stark differences in belief, in personality, and in lifestyle, are an extremely important part of our cosmic significance. It is these differences which enable both human culture, and the Solar-Systic whole, to progress in a desirable direction, through time, as the millennia come and go.

Let us start by answering our questions. The answers might initially strike you as odd, because the purpose of this book is to enable you to come to see the truth of these answers. Such a seeing, such an appreciation, is not your starting point; it is your end destination.

Life, what is it all about?

The Universe strives to come into its optimal state: the non-living Universe strives to become living, and the living Universe strives to become human. The optimal state of the Universe is the human species. Humans following their passions, and fulfilling their potential, is a wonderful thing for themselves, for the planet, and for the Universe. Such a following utilises the amazing abilities that have been provided to us by our Solar System, and it enables the human species to fulfil its cosmic purpose.

Does the existence of the human species matter? In other words, is the human species cosmically significant, or is it just one species among many?

The existence of the human species is immensely cosmically important. The human species is the most valuable and precious life-form that has ever lived, and will ever live, on the Earth.

If the human species is not just one species among many, why is it cosmically significant?

The human species is cosmically significant because it is the saviour of life on Earth. If the human species did not come into existence on the Earth, then the life that has arisen on the Earth would go out of existence.

What does it mean to be cosmically significant?

To talk of a life-form being cosmically significant, is to say that the Universe would be in a vastly inferior state if this life-form did not exist. The human species is cosmically significant. The various species that we call elephants are not cosmically significant. If all the elephants that exist on the Earth were to suddenly die, the Earth, and the Universe, would move to an inferior state. For, the existence of life is preferable to the non-existence of life. However, this move to an inferior state would not be a cosmically significant change, for whilst it is an inferior state, it is not a vastly inferior state.

How does cosmic significance relate to the nature of the Universe? How does the nature of the human species relate to the nature of the Universe?

The Universe is comprised of feelings. Feelings occupy space. The objects that we see around us are collections of such feelings. The more intense a particular collection of feelings is, the more cosmically significant it is. The parts of the Universe that are living are more intense collections of feelings than the parts of the Universe that are non-living. The most intense feelings that exist in the Universe are to be found in the human species. So, the cosmic significance of the human species, the nature of the human species, is directly related to the nature of the Universe. A greater ability to modify is a greater intensity of feeling.

Why has the human species come to dominate the planet?

The human species has come to dominate the planet because humans, like all life-forms, act in accordance with their nature. The nature of life-forms is the nature of the Universe. The non-living Universe was acting in accordance with its nature when it brought forth life. The living Universe was acting in accordance with its nature when it brought forth the human species. The human species acts in accordance with its nature. The activities of the human species arise from its nature. The human species has the same nature as everything else in the Universe; it has come to dominate the planet because it is the epitome of the nature of the Universe. In other words, the fundamental qualities of the human species are the maximal expression of the fundamental nature of the Universe.

Why has the human species acted so as to bring forth a plethora of pressing environmental issues, the most important of which is the phenomenon of human-induced global warming?

The human species has always acted in accordance with its nature, a nature which is the maximal expression of the nature of the Universe. In the past few decades, this expression has escalated and blossomed, and some of the outcomes of this blossoming expression have been labelled as 'environmental issues'. This blossoming expression, along with its associated environmental issues, are an aspect of the cosmic significance of the human species; they are a stepping stone on the journey to the human species fulfilling its cosmic purpose of saving life on Earth. The phenomenon of human-induced global warming is the most important environmental issue. It is simultaneously a sign that life on Earth is thriving, an indicator of immense danger arising from a non-human source, a bringer of impending doom, and the precipitator of the human species fulfilling its cosmic purpose.

It is now time for you to delve into your journey of insight, so that you can eventually come to appreciate the truth of these answers.

Sections

Two Forces

for

Global Warming

Two Forces for Global Warming

Do you know that there are two different forces for global warming? I am sure that you are aware of the phenomenon of human-induced global warming, because we are living through the stage of Solar-Systic unfoldment in which there is inevitably an extremely widespread and increasing awareness of this phenomenon. What you might not be aware of is that human-induced global warming is a force which has come into existence against the backdrop of a pre-existing force for global warming, a force for the warming of the Earth's atmosphere which we can call the force for non-human-induced global warming. It is crucial to appreciate the subtle interconnections between these two forces. For, the force for non-human-induced global warming is a long-standing force, which has set the planetary scene, the planetary conditions, against the backdrop of which the force for human-induced global warming has emerged. Furthermore, the Earth, and our Solar System, are tightly interconnected unfolding wholes, and it is essential to take this into account when considering the emergence of the phenomenon that is human-induced global warming.

It is extremely dangerous to consider the phenomenon of human-induced global warming as an isolated phenomenon, as a human-created problem that is divorced from the wider Solar-Systic context in which it is situated. The force for human-induced global warming, as a non-trivial force, is made possible because of the existence of the force for non-human-induced global warming, and the way that life on Earth has responded to this force. For, a powerful force for human-induced global warming requires the existence of an enormous amount of fossil fuels, and if the force for non-human-induced global warming did not exist, an enormous amount of fossil fuels would not have been created on the Earth. In other words, if it wasn't for the previous existence of the force for non-human-induced global warming, life on Earth would not face a danger from human-induced global warming. Human-induced global warming has to be seen in the context of non-human-induced global warming, and in the context of our unfolding Earth / Solar System, because it is an immensely significant part of this unfolding.

To talk of these two forces for global warming is to talk of two *non-cyclical one-way forces*, forces that grow in strength through time; forces which bring into being an ever-increasing upwards pressure on the temperature of the Earth's atmosphere through time, either as immediate actual pressure or as inevitable pressure in the near future. We are not talking about short-term factors that affect the temperature of the atmosphere in a cyclical fashion, such as the changes in incoming solar radiation that are due to the sunspot cycle of the Sun.

Let us first consider the force for human-induced global warming. Human activities have created a non-cyclical one-way force for global warming. We can imagine a hypothetical scenario in which human activities on the Earth are cyclical in relation to the creation of a force for global warming. To put it simplistically, this would involve fossil fuels being released from underground storage in one time period, in the next time period these fossil fuels would be reburied, and in the following time period they would be released again. This hypothetical cyclical scenario can be compared to reality, the fact that human-induced global warming is a non-cyclical one-way force. Human activities over time have continuously produced a stronger and stronger global warming force. From thousands of years ago to the present, the rainforests of the planet have been progressively destroyed and replaced with agriculture and urban sprawl. Whilst, the history of fossil fuel use is one of ever-increasing transfer from underground storage to the surface of the Earth. The result of these activities is the one-way force that is the force for human-induced global warming.

To be clear, there have always been particular humans and particular communities that live in a global-warming-neutral way. Our interest is not in particular humans/communities; we are interested in the total human impact on the planet. This total human impact is a non-cyclical one-way force for global warming. To talk of human-induced global warming as a non-cyclical one-way force is not to make predictions about the future; it is simply to recognise the reality of the past and the present. It is to say that this force has existed throughout human history and that it still exists today. It might seem to be a bit much to say that this force has existed throughout human history, but this force came into existence when the human species transitioned from a hunter-gatherer lifestyle to a biosphere-transforming agricultural lifestyle. For the vast majority of human history this force has been exceptionally weak, so weak that it hasn't been anything to be worried about, so weak that it didn't given rise to what one would call an 'environmental problem'. The force for human-induced global warming has only become potent since the nineteenth century, due to the bringing forth of technology and the associated release of the stored fossil fuels. It is also important to realise that it is possible for some non-cyclical one-way forces to cease, to become a non-force. When human technology is successfully deployed to actively regulate the atmospheric temperature the force will no longer exist. The total human impact on the planet will then be a non-cyclical force for the sustainability of the atmospheric temperature.

Let us now consider the second type of global warming – the force for non-human-induced global warming. When we consider the entire history of the Earth, from its formation to the present, then the overwhelmingly dominant force exerting upwards pressure on the Earth's atmospheric temperature is the force for non-human-induced global warming. The source of the force for non-human-induced global warming is the increasing amount of solar radiation which reaches the Earth from the Sun. Since life arose on the Earth, 3.5 billion years ago, the amount of solar radiation reaching the Earth from the Sun has increased by a massive 25%. There are short-term cyclical fluctuations in solar radiation which correspond to the sunspot cycle, but these are superimposed onto a continuous long-term one-way upwards process, the forever increasing amount of solar radiation being propelled towards the Earth. This one-way process gives rise to the force that is non-human-induced global warming.

Now is a good time to stress the crucial difference between the existence of 'a force for global warming' and 'the actuality of global warming'. It is easy to focus on 'the actuality of global warming' – measuring atmospheric temperatures, and observed effects such as sea-level rises and extreme weather events. Measuring atmospheric temperatures, and taking note of the effects of increased atmospheric temperatures, is an important endeavour. However, what is immensely more important is to appreciate the existence, nature and strength of 'the forces for global warming'. This is essential because we want to understand why the actual global warming that has occurred has occurred, and we want to understand the immensity of the forces for global warming that currently exist that have not yet manifested into actual global warming.

A force for global warming could, as soon as it comes into existence, lead to immediate actual global warming. However, it is much more likely that when a force for global warming comes into existence, that a chain of counterbalancing reactions are set in place, as this force is attempted to be offset, in order to prevent actual global warming from occurring. For, life on Earth does everything that it possibly can to prevent its atmospheric temperature from rising. Whilst actual global warming can in this way be averted, it is important to realise that there are two different types of counterbalancing reactions. The first type of reaction provides a sustainable long-term force for global cooling. Whilst, the second type of reaction provides a temporary respite, a short-term prevention of actual global warming which brings with it the creation of a force for future global warming.

The first type of counterbalancing reaction is a sustainable long-term force for global cooling because it removes greenhouse gases from the surface of the Earth, and moves them into long-term storage under the Earth's surface. To be clear, to talk of 'the surface of the Earth', is to talk of the Earth's atmosphere, its oceans and its land surfaces; greenhouse gases can easily move between the various components of 'the surface of the Earth'. If greenhouse gases move to 'long-term storage under the Earth's surface', then they are removed from 'the surface of the Earth'. These counterbalancing reactions occur over long timescales, we are talking about hundreds of thousands, and millions, of years. These counterbalancing reactions are closely associated with the force for non-human-induced global warming.

Let us now consider the second type of counterbalancing reaction. This reaction provides a short-term prevention of actual global warming by temporarily pulling greenhouse gases from the atmosphere into the depths of the ocean. This counterbalancing is only temporary because the oceanic cycles that we are concerned with here operate over a short time period. This counterbalancing reaction involves the creation of a force for future global warming; it creates a temporary masking of the original force for global warming, by temporarily delaying the final atmospheric global warming impacts of the original force. In this way, an increasingly potent force for global warming can, and has, come into existence. Such a situation is obviously deeply worrying, because the increasing potency of the force is effectively and increasingly masked, waiting to be unleashed. The greater the force becomes, the greater the actual global warming that is set to occur when the force is unleashed. These counterbalancing reactions occur over timescales of several hundred years, and they are particularly significant when it comes to getting to know the force for human-induced global warming.

Coming to know 'the forces for global warming' is obviously immensely more important than recording 'the actuality of global warming'. You will come to appreciate that if the forces for global warming that currently exist on the Earth were an iceberg, that the actuality of global warming that has occurred would equate to a tiny part of the tip of this iceberg. Furthermore, given that life on Earth does all that it possibly can to prevent its atmospheric temperature from rising, you can appreciate that when this counterbalancing starts to fail, when life on Earth has run out of options and the atmospheric temperature starts to rise, that this means that life on Earth is losing its battle for survival. Yet, perhaps seemingly paradoxically, you will also come to realise that an initial increase in atmospheric temperature caused by the force for human-induced global warming means that life on Earth is winning its battle for survival. You will come to know the immense forces for global warming that currently exist, immense forces which are set to overwhelm the planet in the coming centuries, causing the Earth's atmospheric temperature to rocket to a level that is forever too hot for life to survive and thrive.

If we weren't in such a perilous situation, one could find it comical that the deluded human species believes that the appropriate response to the global warming situation that it currently faces is to attempt to reduce its emissions of fossil fuels. One could easily laugh if one was observing the human species attempt to extinguish a 2000 hectare blazing inferno by using the water that is in a child's water pistol. After all, they could have utilised their strengths, they could have deployed fire engines and fire fighting turbines; they could have deployed planes and helicopters to smother the flames through dropping enormous quantities of water and fire-retardant chemicals. What is our lesson? Continued delusion leads to death and suffering; whereas, a future in which life survives and thrives will come into being if the human species maximally utilises its skills, talents and abilities. The deluded encourage us to deny our nature; our nature is to explore, to modify, to appropriate resources, to dominate, to control. The deluded want us to respond to the immense force for global warming that is upon us by attempting to reduce our fossil fuel emissions. What a lame, ineffectual and feeble response!

We need to consider what it means for there to be sustainability in terms of the Earth's atmospheric temperature:

Sustainability of the atmospheric temperature either requires the absence of any forces for global warming, or the continuous long-term counterbalancing of forces for global warming with forces for global cooling.

The former possibility for sustainability is not an option because the force for non-human-induced global warming of the Earth's atmosphere will exist as long as the Earth exists. So, sustainability requires a force for global cooling to continuously counterbalance the forces for global warming. We need to delve into the history of the Earth's atmosphere so that we can appreciate how the interplay between two opposing forces – the force for non-human-induced global warming and a force for global cooling – has resulted in a sustainable atmospheric temperature, a temperature that has always been within the range that life requires in order to survive and thrive. When we have come to know this historical interplay, this epic battle, this long-term counterbalancing, we will then be able to clearly see why the human species is the precious saviour of the life that has arisen on the Earth. Life on Earth brought us into being so that it could save itself, through enabling the continued counterbalancing of the force for non-human-induced global warming, with a desperately needed new force for global cooling. Through the human species, life on Earth is evolving a suit of technological armour.

The 25% increase in incoming solar radiation since life arose on the planet is an immensely powerful force for global warming. Yet, despite the immensity of this force for non-human-induced global warming, the temperature of the Earth's atmosphere hasn't increased. This is because this force for global warming has been counterbalanced by a force for global cooling, thereby preventing the force becoming an actuality. If this force for global cooling did not exist, the Earth's atmospheric temperature would now be way too hot for complex life-forms to exist; the Earth would today be a barren place. It is essential to fully appreciate the nature of this counterbalancing force for global cooling; for, despite the long duration of its existence, it is a time limited force; and, it has already approached its expiration.

We need to envision the past and the future if we are to appreciate the present. In the past, the amount of incoming solar radiation steadily increased over time; it is currently 25% higher than when life arose on the Earth. In the future, the amount of incoming solar radiation will continue to increase as the Sun continues to age and ultimately expire. It is important to envision this, because it is part of the bigger picture; it is the backdrop to our evolution and to the bringing forth of the force for human-induced global warming. Seeing the bigger picture enables us to appreciate the forces for global warming that exist in the present, and such an appreciation is a prerequisite for understanding what is set to occur in the coming decades and centuries.

We know that life on Earth needs particular conditions in order to survive and thrive; in particular, it needs an atmospheric temperature of between 10°C and 20°C. Since life arose on the Earth the average temperature of the atmosphere (the Global Mean Surface Temperature / GMST) has always been in this range. The reason that it has not moved above 20°C is that life on Earth has deployed a force for global cooling in order to keep the atmospheric temperature favourable for its continued thriving. There are five stages of inevitable progression which the Earth is transitioning through. The first stage occurred when life arose on the Earth; at this time the amount of solar radiation reaching the planet was at the low end of what life requires; so, life had to bide its time, and wait for the incoming solar radiation to increase, thereby creating the conditions that it needed to thrive. The second stage occurred when the level of incoming solar radiation was ideal for the survival and thriving of life. At this time life thrived and had no need to deploy a force for global cooling. The third stage involves the level of incoming solar radiation being too high for the survival and thriving of life. In this stage, in order to continue to survive and thrive, life on Earth had to offset this unwelcomely high level of solar radiation by creating a force for global cooling. The fourth stage, which started about two million years ago, involves the gradual breakdown of the force for global cooling. The force for global cooling deployed by life on Earth was always going to be time limited; it could not go on forever. In this fourth stage, life on Earth struggles to keep the Earth's atmospheric temperature down, it starts to lose its battle with the force for non-human-induced global warming, and it strives to bring forth its technological saviour species before it is too late. The fifth stage, which we are about to enter, involves the bringing forth by life on Earth of a new force for global cooling, a technological force for global cooling, via the human species.

Our objective is to explore this journey from the third stage, to the fourth stage, to the fifth stage. In the third stage, life on Earth deploys a force for global cooling in order to keep the atmospheric temperature down to a level that enables its complex life-forms to exist, to thrive, and to evolve. What is the nature of this counterbalancing force? As the incoming solar radiation has increased, life on Earth has responded by pulling an increasing amount of carbon dioxide out of its atmosphere and into itself, thereby forming carbon, and this has subsequently been transported into storage under the surface of the Earth. Carbon dioxide is a 'greenhouse gas'. Fewer greenhouse gases in the atmosphere means a smaller 'greenhouse effect' and therefore a lower GMST. The fact that this transfer is from the atmosphere to underground storage means that this is a continuous long-term sustainable counterbalancing. If the carbon dioxide was simply moved from the atmosphere to the biogeochemical cycles that occur at the surface of the Earth, and didn't make it to underground storage, then, within a very short period of time, the carbon dioxide would make it back into the atmosphere; so, this would be a short-term unsustainable counterbalancing.

The increasing transfer of carbon dioxide from the Earth's atmosphere into itself, where it becomes carbon, and the subsequent movement of this carbon to storage under the Earth's surface, is an extremely powerful force for global cooling which life on Earth has deployed in order to counterbalance the force for non-human-induced global warming. This homeostatic regulation to maintain the conditions that life on Earth needs in order to survive and thrive can be successful for an exceptionally long period of time. This is because when life on Earth entered stage three – when it needed to create a force for global cooling – there was an enormous amount of carbon dioxide in the Earth's atmosphere which could be removed. The carbon dioxide that life has drawn into itself has ultimately formed fossil fuels – coal, gas and oil – and chalk. As the incoming solar radiation gradually increased, life on Earth sensed this increase and reacted by increasing its activity, thereby pulling an increasing amount of carbon dioxide out of its atmosphere and into itself. The increasing movement of this carbon to underground storage constitutes the counterbalancing force for global cooling that has been deployed by life on Earth. This force for global cooling has provided a favourable and sustainable atmospheric temperature for an immensely long period of time.

The fourth stage, which started about two million years ago, involves the gradual breakdown of the force for global cooling. This force is powered by the removal of carbon dioxide from the Earth's atmosphere. This removal is straightforward when there is an abundance of carbon dioxide in the atmosphere. However, as atmospheric carbon dioxide levels gradually fall over time, the point eventually comes at which this removal becomes difficult, and as time passes it eventually becomes impossible. In other words, the force for global cooling starts to breakdown, then it expires. A thriving life-bearing planet needs to have carbon dioxide in its atmosphere; it can only fall to a certain level. When this level is reached, as it is in the fourth stage, life on Earth is in serious peril. The weakening of the force for global cooling means that the Earth's atmospheric temperature could easily shoot upwards due to being overwhelmed by the increasing strength of the force for non-human-induced global warming. The gradual breakdown of the force for global cooling is revealed by the transitions that have been occurring between glacials and interglacials, in accordance with Milankovich Cycles. These transitions are an indicator of increasing planetary sickness; they are a sign that life on Earth is losing its battle with the force for non-human-induced global warming.

We have, so far, been considering the interactions of the forces which existed on the Earth before the human species evolved. We need to take a moment here. It is essential to realize that when the human species evolved, the Earth was not a harmonious peaceful place. Such a realization is of the utmost importance! There had been a great battle ongoing for millennia upon millennia, epoch upon epoch. Two immense opposing forces had been locking horns against each other. The Sun, sending out ever-increasing solar radiation; life on Earth, retaliating, by removing carbon dioxide from its atmosphere. This is an on-going battle for the atmosphere, an on-going battle for the survival of life on Earth. At first glance, this battle appears to be between enemies of vastly unequal power – an epic 'David versus Goliath' confrontation. The Sun is Goliath; it is unrepentant in its ability to send forever increasing amounts of solar energy to the Earth. Life on Earth is David; it is an inferior opponent, its weapons are weaker, it is smaller and more fragile; yet it is seemingly determined to be victorious. Life on Earth is still resisting the increasing solar radiation today despite the increasing difficulty of its task. And there is hope on the horizon. For, whilst it was facing obliteration, life on Earth has continued the battle for long enough to enable it to bring forth technology, via the bringing into being of the human species. Life on Earth may yet overcome the odds and be victorious. If life on Earth is able to deploy technology to form a new force for global cooling, then the life that has arisen on the Earth will have a long and glorious future.

Life on Earth will be victorious. We are about to enter the fifth stage, which involves the bringing forth by life on Earth of a new force for global cooling, a technological force for global cooling, via the activities of the human species. How will this come about? And, what about the force for human-induced global warming; how does this fit into the bigger picture? In order to address these questions we need to delve into an exploration of the evolution and activities of the human species.

Life on Earth became technological at the moment in time in the evolution of the Earth / Solar System that it needed to become technological. How amazing is that! The human species, the bringer forth of technology, was brought into existence at the time when life on Earth's counterbalancing force for global cooling, the force that has been making the Earth a habitable place for life, was breaking down and needed a technological replacement. The coming into existence of technology on the Earth heralds the prospect of a bright future. For, technology can be deployed as a new counterbalancing force for global cooling, a desperately needed new weapon in life on Earth's armoury as it seeks to win its battle with the Sun.

However, things aren't all rosy. Things aren't quite this simple. For, whilst the bringing forth of technology heralds the prospect of a bright future, it also brings with it imminent danger. The bringing forth of technology ushers the near future forward to the present. The bringing forth of technology brings things to a head. The bringing forth of technology is a watershed moment on the Earth. For, whilst the bringing forth of technology facilitates a long-term future for life on Earth, its coming into being brings with it a period of disruption which actually threatens the existence of life on Earth. So, whilst life on Earth was in peril before the coming into existence of the human species / technology; this peril was not a dire emergency. The bringing into existence of technology brings the peril forward from the near future to the present. We thus face a dire emergency in the upcoming decades and centuries. Furthermore, the bringing forth of technology entails an epoch of delusion, in which the bringer forth of technology believes that it is the destroyer of life on its planet; at the time that it is bringing forth technology, it is unable to see that it is the technological saviour of life on Earth. This raises a question: Given its delusion, how does the human species come to embrace and fulfil its cosmic purpose, and deploy the required technology? One possible answer to this question can be found in this section: *The "Mysterious Prince"*. For now, let us start to explore the human presence on the planet.

The coming into existence of technology requires life to become analytical and scientific. Technology is the zenith of the ability of life to transform, modify, rationalise and investigate. That which brings technology into being has to see itself as non-natural; it has to see the world around it as resources for its use; it has to be deluded to both the nature of the Universe and its cosmic purpose. For, the bringing forth of technology requires life to impart immense suffering and pain onto itself. If the human species could truly see the nature of its fellow planetary life-forms, if the human species could see how similar it is to all of these diverse life-forms, then it could not possibly bring itself to utilise them in the way that it has. Mass vivisection; mass agriculture; mass enslavement; mass slaughter! If the human species was fully aware of what it was doing to its fellow brothers and sisters these actions would have driven it to insanity! Such actions require the perpetrator to be unaware of what they are doing. The bringing forth of technology, the powering forward of human culture in order to save life on Earth, requires the bringer forth of technology to be deluded, to see everything as mere resources which it can utilise without a second thought.

Those few members of the human species who are tuned in to the nature of the Universe, and the nature of life, shy away from dominating and transforming the world around them. This means that the survival of life on Earth requires the dominating/modifying activities of the deluded masses.

The human species cannot repudiate its nature, its impulse to transform, modify and utilise. The bringing forth of technology entails planetary transformation; it entails the wholesale marshalling of the Earth's resources in order to feed the growth and the complexification of technology. Part of this marshalling of resources is the inevitable utilisation of the fossil fuels that are to be found under the Earth's surface. When the human species discovers this immense treasure trove of energy it cannot believe its luck. What an amazing resource that can be used to power forward the development and complexification of technology! At the time that these fossil fuels are released from underground storage and burned, there is no knowledge, no awareness, of the reason why the carbon was stored underground in the first place. There is thus no realisation of the immense magnitude of the deadly storm that is being unleashed, as these fossil fuels are rapidly and excitedly released from safe storage.

Releasing the stored fossil fuels obviously has potentially catastrophic consequences. Fossil fuels were created for a reason, for a purpose. They are created from the carbon dioxide that was removed from the atmosphere in order to keep the atmospheric temperature favourable for life. Their storage was counterbalancing the force for non-human-induced global warming. Their release from storage therefore heralds an upcoming storm of immense proportions. Not an immediate storm. The storm is yet to come! For, the release of the stored fossil fuels has, so far, led to only a small amount of global warming. As the human species has burned fossil fuels atmospheric carbon dioxide levels have increased, this has enhanced the 'greenhouse effect' and the Earth's atmospheric temperature has consequently nudged upwards. The storm that is about to hit us comes not from the carbon dioxide that is currently in the atmosphere, but from the carbon dioxide that has been increasingly accumulating in the oceans. This build-up has created an immense force for future global warming. The wholesale burning of fossil fuels started with the arrival of the Industrial Revolution. When these fossil fuels were burned, about half of the released carbon dioxide was immediately sucked into the ocean where it entered the deep ocean thermohaline circulation. And the same thing has been happening ever since. Think of all of the fossil fuels that have been burned across the world since the start of the Industrial Revolution; what a colossal amount this is! About half of the carbon dioxide released from this burning was sucked into the thermohaline circulation, and the overwhelming majority of it is still there.

The thermohaline circulation is a process which involves dense cold water sinking at high latitudes and travelling very slowly through the ocean depths until it eventually reaches the northern Indian Ocean and the northern Pacific Ocean where it resurfaces. When carbon dioxide sinks into the thermohaline circulation it takes from 100 years to 1000 years to re-enter the atmosphere. This means that almost all of the ginormous amount of carbon dioxide that has become stored in the thermohaline circulation since the start of the Industrial Revolution has yet to re-enter the atmosphere. This travelling carbon dioxide is a massive force for future global warming. When this carbon dioxide starts to be released *en masse,* in the upcoming centuries, the temperature of the Earth's atmosphere is set to shoot upwards. By the year 3000 – when all of the carbon dioxide that is currently stored in the thermohaline has been released – the Earth's atmospheric temperature could very easily be too hot for humans and for other complex life-forms.

The likelihood of a climatic tipping point being passed, taking us to a much hotter planet, is enhanced by the fact that the world that the long-travelling oceanic carbon dioxide will be gushing out into is a very different world to the world that it departed from. In this vastly different world, neither the land surfaces nor the oceans are such a welcoming home for atmospheric carbon dioxide. Oh no! When the long-travelling oceanic carbon dioxide gushes out of the thermohaline circulation it will accumulate in the Earth's atmosphere. Atmospheric carbon dioxide concentrations will spiral upwards. Why is the world to which the long-travelling oceanic carbon dioxide will be returning to so different? Why is it a vastly different world to the world that it departed from?

Let us first consider the transformation of the Earth's land surfaces. Since the start of the Industrial Revolution the human species has transformed the Earth's land surfaces through mass deforestation. The extent and speed of the deforestation that has occurred in this period is mind-boggling. A ginormous amount of lush rainforest (which is a large and immensely valuable 'carbon sink') has been chopped down by chainsaws, and has been replaced by urban sprawl, concrete, industrial agriculture, mining operations and roads. Such a transformation makes it harder for life on Earth to keep its atmospheric temperature down.

Let us now turn to the transformation of the Earth's oceans. Since the start of the Industrial Revolution the Earth's atmospheric temperature has increased by approximately one degree Celsius. This warming of the atmosphere, caused by the burning of fossil fuels, has caused the oceans to warm. This is of great importance because carbon dioxide is less soluble in warmer water. This warming of the atmosphere has also caused the melting of icebergs and glaciers in the North Atlantic and Greenland. This is of great importance because as the continuously melting freshwater pours into the Atlantic Ocean it slows down the thermohaline circulation. And a slower thermohaline circulation means less carbon dioxide in the oceans and more carbon dioxide in the atmosphere.

So, at the start of the Industrial Revolution, when the long-travelling oceanic carbon dioxide that we are exploring – that which emanated from the mass burning of fossil fuels – started to enter the thermohaline circulation, the Earth's land surfaces contained immense lush rainforests, and the thermohaline circulation was vigorous. The rainforests were an immense 'carbon sink', and the thermohaline circulation was vigorously sucking in enormous amounts of carbon dioxide. One event – the Industrial Revolution – brought this world to an end and ushered in a vastly different world. This one event led to a transformation of the Earth's land surfaces through mass deforestation. This one event also simultaneously caused an enormous amount of carbon dioxide to build up in the thermohaline circulation, whilst also initiating a chain of events which are causing the thermohaline circulation to increasingly slow down! This one event also caused the oceans to be a less conducive home for carbon dioxide through raising the oceanic temperature. All of this, the coming into existence of this vastly changed world, means that when the long-travelling oceanic carbon dioxide starts to gush out of the oceans it will start accumulating in the Earth's atmosphere. Both the land surfaces, and the oceans, have been transformed since the long-travelling oceanic carbon dioxide departed. A perfect storm is about to hit the Earth in the coming decades and centuries.

We can see that atmospheric carbon dioxide concentrations are set to spiral upwards for centuries. The Earth's atmospheric temperature is set to shoot upwards. Tipping points are set to be passed. Large-scale discontinuities, such as a runaway greenhouse effect resulting from the destabilisation of methane clathrate reservoirs, are set to be triggered. We are on the verge of the Earth becoming an inhospitable and barren place. At this point, it is worth repeating what we have already said:

> *Whilst the bringing forth of technology heralds the prospect of a bright future, it also brings with it imminent danger. The bringing forth of technology ushers the near future forward to the present. The bringing forth of technology brings things to a head. The bringing forth of technology is a watershed moment on the Earth. For, whilst the bringing forth of technology facilitates a long-term future for life on Earth, its coming into being brings with it a period of disruption which actually threatens the existence of life on Earth. So, whilst life on Earth was in peril before the coming into existence of the human species / technology; this peril was not a dire emergency. The bringing into existence of technology brings the peril forward from the near future to the present. We thus face a dire emergency in the upcoming decades and centuries.*

It is crucial to realise that we cannot stop the perfect storm that is about to hit the Earth; it has been created in the past and it will be unleashed onto the Earth in the coming decades and centuries. What we need to do is to come face-to-face with the perfect storm, to acknowledge it, to passionately embrace it, and to adequately prepare for its imminent arrival with an adequately powerful response. Let us be clear – the perfect storm wouldn't miraculously go away if the human species were to stop using fossil fuels today. Reducing greenhouse gas emissions, reducing fossil fuel use, will not do anything to avert the imminent danger that is hurtling towards life on Earth. Given the overall situation that we are currently in, such reductions are a trifling insignificance. Such reductions are not a solution to the problem that we face. More than this, the attempt is a harmful distraction that will result in avoidable suffering and death. The attempt to address the global warming situation that life on Earth currently faces through fossil fuel emission reduction efforts is not just ineffectual, a total waste of time and resources, it is positively harmful; for, it is distracting us, and preventing us from ploughing our energies, and our resources, into the only possible solution to the global warming situation that we currently face.

We are currently living through a one-off transformation, a spectacular metamorphosis that will only ever happen once on the Earth, as the effects of the bringing forth of technology pervade the planet and bring the situation to a head. After the coming centuries have passed there are two possible futures. Either technology will have taken control of the Earth's atmospheric temperature and the life that has arisen on the Earth will have a long and glorious future, or all will be lost and life will be decimated. The bringing of things to a head, this one-off event, can be thought of as the mother of all perfect storms. You know that life on Earth was in peril before the human species came into existence. Indeed, to overcome this peril, through bringing into being a technological force for global cooling, is the reason why the human species was brought forth into existence. Yet, it is the very activities that are involved in bringing forth technology, technology that is needed to save life on Earth, that have brought the future peril that life on Earth was facing into the present, thereby giving rise to the perfect storm that is about to hit the Earth.

The predominant part of the perfect storm is the progressive breakdown of the force for global cooling. The low levels of carbon dioxide in the Earth's atmosphere mean that life on Earth is unable to continue to effortlessly remove carbon dioxide from its atmosphere and move it to underground storage. The coming into being of the perfect storm, in its totality, is triggered by the bringing forth of technology, which involves the human release of fossil fuels from underground storage, in tandem with the transformation of the surface of the Earth through mass deforestation, urbanisation and industrial agriculture. Another aspect of this perfect storm is that the bringing forth of science and technology enables the human population size to escalate, due to enabling humans to live longer, and due to the fruits of industrial agriculture. This escalating human population also has a tremendous thirst for the new consumer goods that have been produced, and for the travel that technology enables. It cannot help itself; it is a slave to this unquenchable thirst! Furthermore, whilst the richest humans use the most resources, other humans across the world, and poorer countries across the world, aspire to be as rich and as resource-intensive in their lifestyles as the richest. Another aspect of this perfect storm is that when the released fossil fuels are burned, a large proportion of them make their way into a temporary home in the thermohaline circulation, creating a massive force for future global warming. This force is set to turn into significant actual global warming in the coming decades and centuries, as the long-travelling oceanic carbon dioxide makes its dramatic return into a vastly changed world. These things are all aspects of the perfect storm that is an immediate precursor to life on a planet bringing forth its technological force for global cooling.

The immensity of this currently existing perfect storm of forces for global warming needs to be appreciated. For, such an appreciation is needed in order to clearly see why focusing our energies on fossil fuel emission reduction efforts is totally misplaced. The pursuit of such a strategy will just result in undesirable outcomes, such as immense amounts of suffering and death that could have been avoided. Nothing good will come from our attempt to deal with the global warming situation that we face through limiting our fossil fuel emissions. The perfect storm already exists and it is set to be unleashed.

Whilst facing up to the existence of this perfect storm, we need to keep in mind that human actions, human activities, the bringing forth of technology, the release and burning of the previously stored fossil fuels, are all an expression of the fundamental nature of the Universe. All of these things are an important part of the cosmic unfolding of life on Earth. Our cosmic purpose as a species was to bring this moment into being, to bring forth technology, to deploy technology to control the temperature of the Earth's atmosphere in order to enable the continued counterbalancing of the force for non-human-induced global warming. We are the part of life on Earth that is the saviour of the life that has arisen on the Earth.

We cannot shirk our responsibility. We cannot hide away and evade our nature. If we attempt to deny our nature, and to immediately slash our fossil fuel use, we would continue to fail.

And even if we were to succeed in this endeavour, that would also be a failure; it is the wrong objective; it is an objective that cannot possibly achieve its desired outcome, which is to provide a stable and sustainable atmospheric temperature that is conducive to the continued thriving of the life-forms that reside on the Earth.

We need to come face-to-face with our cosmic purpose, with our responsibility to save life on Earth, and get to work. The more we dither and delay, the greater the extent of the global warming that will occur, and the greater will be the unnecessary suffering and death to the life-forms of the Earth.

This is perhaps quite a lot for one to get one's head around. It might be helpful to partially repeat things, and to express them in a slightly different way. For hundreds of millions of years life on Earth has been successfully regulating the Earth's atmospheric temperature. It has achieved this feat through counterbalancing increasing incoming solar radiation by drawing carbon dioxide down from the Earth's atmosphere and into itself, forming carbon, and then moving this into safe storage under the Earth's surface. The time would always inevitably come when this counterbalancing would stop working, due to the lack of possible carbon dioxide to drawdown in this way. Life on Earth always needed to progress to the point of becoming technological if it was to survive. Becoming technological takes an immensely long time. By the time that life on Earth was able to become technological it was inevitable that its non-technological ability to counterbalance the increasing solar radiation would already be at the point of imminent collapse. When life becomes technological it causes massive temporary disruption through releasing the stored fossil fuels and transforming the face of its planet.

The planet that existed before technology came into being is never to be returned to. Things have moved on and a completely different world exists. The epoch of the bringing forth of technology brings things to a head: the previous slow demise of the ability of life on Earth to counterbalance increasing incoming solar radiation, is now replaced by an urgent need for its replacement with a technological substitute.

The human species has to realise all of this and embrace its nature and its cosmic purpose. There is no going backwards. There is no point hankering for a past planet of hypothesised harmony and sustainability. There never was such a thing. For hundreds of millions of years there has been a colossal collision of forces, the battle for the atmosphere that we have been considering, between the force for non-human-induced global warming and the attempts of life on Earth to counterbalance this force with its own force for global cooling. The time has come in the evolution of our Earth / Solar System where this force needs to be replaced by the force of technology. Such a weapon is life on Earth's last throw of the dice, its last hope for survival. It is our mission to embrace our destiny and thereby enable life on Earth to thrive into the far distant future, through our ability to take control of the Earth's atmospheric temperature.

You might have been thinking to yourself that the human release of fossil fuels could provide a boost to the beleaguered counterbalancing ability of life on Earth. For, you might think that it provides an additional opportunity for these released greenhouse gases to be drawn down again in the same manner that they were before. But, alas, this is a forlorn hope. The world is now a very different world to the world of the past. Everything has changed and things have come to a head. The drawdown mechanisms of the past are slow, and the need for action is now urgent. The human release of the stored fossil fuels has not created a problem, it has just brought the pre-existing problem to a head, brought it forward from the very near future to the present.

Far from being a problem, the human release of the stored fossil fuels is part of the solution to the pre-existing problem; for, it is an indicator that life on Earth is thriving, that life on Earth has become technological, that life on Earth is well on its way to saving itself. We just need to embrace our destiny and complete the job.

The hard work has already been done; the hard work was the bringing into existence of the human species. We need to become aware of our destiny, of our cosmic purpose, of the realities of our nature and of our unfolding Earth / Solar System. The force for human-induced global warming is part of the solution to the main issue that life on Earth faces, which is the need to counterbalance the force for non-human-induced global warming.

It is important to realise that if the only force for global warming was human-induced global warming, then simply reversing what we have done, through drawing back down the fossil fuels that we have released from underground storage, and re-storing them, would be the end of the story. But this isn't the end of the story; it is just the beginning. The fundamental force for global warming is non-human-induced global warming, and counterbalancing this force requires more than technologies that regulate greenhouse gas concentrations in the atmosphere. We are now rapidly approaching the moment in Solar-Systic evolution at which counterbalancing the force for non-human-induced global warming requires technologies that block incoming solar radiation, thereby preventing it from reaching the Earth's atmosphere.

The situation that we are currently in, the perfect storm that we face, is a delicate interplay between a multitude of forces: the force for non-human-induced global warming, the force for human-induced global warming, the planetary transformations and temporary perturbations which have occurred as technology has been brought forth, the breakdown in life on Earth's ability to counterbalance the force for non-human-induced global warming with its force for global cooling, and the unquenchable force that is constituted out of the expression of the skills, abilities, passions and talents of the human species. Now that all of these forces have revealed themselves to us, the only conclusion that we can reach is that our future, the future thriving of life on Earth, and even the future survival of the life that has arisen on the Earth, requires us to embrace our destiny and deploy technology to regulate the temperature of the Earth's atmosphere. There is no time to waste. There is no time to faff around in the delusional belief that reducing fossil fuel emissions is a solution to the situation that we are in.

It is time to wake up and embrace our cosmic purpose!

Life on Earth is depending on us!

Life on Earth is willing us to succeed!

Global Warming

&

Human Cosmic Purpose

"What is man? – so I might begin; how does it happen that the world contains such a thing, which ferments like a chaos or moulders like a rotten tree, and never grows to ripeness? How can Nature tolerate this sour grape among her sweet clusters?" [1]

Friedrich Hölderlin

[1]Hölderlin, F. (1797) ' Hyperion' in Eric L. Santner (ed.), *Hyperion and Selected Poems,* New York: Continuum, 1990, p. 35.

Dear Friedrich

The human species is the zenith of the evolutionary progression of life on Earth. That which troubles you, the chaos and the rottenness, enables the bringing into being of wondrous life-giving fruits. The need for these joyous delicacies explains the tolerance of the sourness that precedes them. Rest assured my friend, the human species will reach ripeness, but first it needs to fulfil its cosmic purpose!

'Save the planet'

I wonder how many times that you have been urged by someone – a friend, a celebrity, a politician, an environmentalist – to change the way that you live in order to help to 'save the planet'?

Such urgings are commonplace in our current stage of Solar-Systic unfoldment, the 'guilt-tinged' age of the technological explosion. Such urgings currently pervade the media, popular culture and academia. Such urgings are central to the way that we have come to see ourselves as a species. As the age of the technological explosion has advanced these urgings have escalated; we are now faced with a pandemic of urgings to 'save the planet'.

But, does the planet need to be saved?

'Save life on Earth'

The planet Earth does not need saving; those who urge us to change the way that we live know this. When they talk of the need to 'save the planet', what they really mean is that they believe that there is a need to save the life that resides on the Earth.

Some people believe that life on Earth in its totality needs to be saved; in other words, they believe that the Earth could soon become a barren and lifeless place. Other people believe that it is only part of life on Earth that needs to be saved; in other words, they believe that a certain proportion of the species/life-forms that currently exist need to be saved, but that if this proportion of life-forms were to all die that the Earth would still be home to a plethora of wonderful life-forms.

The planet Earth itself is exceptionally robust; there is no sense in which it needs to be saved.

But, does the life that resides on the Earth need to be saved?

What is it that life on Earth supposedly needs to be saved from?

Those people who urge us to 'save the planet' / 'save life on Earth' do so because they believe that life on Earth needs to be saved due to widespread perturbations and destabilisations that have been caused by the activities of the human species. They believe that these widespread perturbations and destabilisations pose a threat to the continued thriving, even the continued survival, of life on Earth.

In other words, the human species has come to conceptualise itself as an entity that poses a threat to the continued thriving of life on Earth, due to the magnitude of its collective impact on the Earth's biosphere. The current pandemic of urgings to 'save the planet', are urgings for the human species to rein in its destabilising activity, in order to avert its supposed potentially catastrophic impact on the Earth's biosphere. The ethos underlying this conceptualisation, and its associated pandemic of urgings, is clearly that the interests of life on Earth are best served by the human species 'getting out of the way' and 'leaving things to nature' as much as possible.

In short, the pandemic of urgings has arisen due to the belief that life on Earth needs to be saved from the destructive activities of the human species.

Saving ourselves

When some people urge us to 'save the planet' what they really mean is that they believe that the human species needs to be saved; they believe that the human species needs to change the way that it collectively acts in order to save itself.

To believe such a thing is not to deny that the perturbations and destabilisations caused by the human species can bring suffering and death to non-human life-forms. One can accept that this is the case whilst also believing that, fundamentally, it is the human species that needs to be saved. These people believe that life on Earth was thriving *before* the human species came into existence, that the perturbations and destabilisations brought into being by the human species might lead to the extinction of both the human species and some other species, but that if this eventuality came to pass that life on Earth would be fine and dandy and would be positively thriving in the distant future. According to this way of thinking, whilst the human species, and some other species, would no longer exist, they would have been replaced by other wonderfully complex and interesting life-forms and life on Earth would be flourishing.

To talk of 'saving ourselves' is, as we have been exploring, usually to talk of preventing the extinction of the human species. However, one could use the phrase in a less alarmist way, and take it to mean 'saving human civilisation in its current form'. For example, if sea levels were to spiral upwards causing most of the world's cities to become submerged into oblivion, then the human species wouldn't become extinct, but human civilisation would be very different to the way that it is now.

In short, some of the people who are urging us to change the way that we live do so because they believe that we need to save ourselves. They believe that life on Earth, as a totality, as a flourishing entity, does not need to be saved, but that the human species does need to be saved.

The trap of the flawed extrapolation

We have seen that some people believe that we need to change the way that we live in order to save ourselves, whilst also believing that life on Earth would have a long and glorious future if we were to go extinct. This isn't true; such a belief is wrong. This erroneous belief derives from a flawed extrapolation from the past to the future.

Life on Earth was thriving for millions of years before the human species came into existence, so it is very easy to fall into the trap of believing that it would continue to thrive in the future if the human species became extinct. In the past, life on Earth has encountered serious setbacks, what we call 'mass extinction' episodes. Following each of these past episodes life on Earth has bounced back to a level of health that has superseded that which existed prior to the episode.

We are in the midst of a human-induced mass extinction episode, but the past is currently not a good guide to the future in terms of whether life would bounce back and thrive following a mass extinction episode which included the extinction of the human species. Such an extinction episode, at our current stage of Solar-Systic unfoldment, would propel us down the path that leads to life on Earth being forever decimated. If one believes that following such an episode life would bounce back and thrive, like it has in the past, then one has fallen into the trap of the flawed extrapolation.

Our mission

The human species does need to be saved; there is no doubt about that. However, our mission, throughout the rest of this section, is to come to appreciate why it is not just the human species that needs to be saved; rather, it is the broader phenomenon of life on Earth that needs to be saved. When one has acquired this appreciation, then one will have escaped the clutches of the trap of the flawed extrapolation.

The phrase 'life on Earth' refers to that part of the Earth that is living, as opposed to non-living. Life on Earth can be absolutely thriving, like it is currently, being constituted out of a ginormous plethora of wonderfully complex life-forms. Life on Earth can also be in a terrible state, containing nothing more than simple single-celled organisms.

To save life on Earth is to maintain the state of thriving. If life on Earth exists in the future, but is in a terrible state, then all is lost; life on Earth would be close to lifeless and would be lurching headlong into oblivion.

Let us explore why life on Earth needs to be saved. And, let us come to understand why the main cause of the need for saviour has nothing to do with the human species.

The main threat to life on Earth

There is no doubt that life on Earth needs to be saved. The cause of the need for saviour is that our Solar System has unfolded to the point at which the ability of life on Earth to continue to keep atmospheric conditions favourable for its continued thriving, through the removal of carbon dioxide from its atmosphere, has almost ceased.

Atmospheric carbon dioxide concentrations on the Earth used to be 5000 parts per million (ppm). That is not a typing mistake, I didn't mean to say five hundred ppm; there used to be five thousand ppm of carbon dioxide in the Earth's atmosphere! Where has all that carbon dioxide gone?

As our Solar System has unfolded the amount of solar radiation that has been propelled to the Earth from the Sun has continuously increased. This continuous increasing is a force for the warming of the Earth's atmospheric temperature. If this force wasn't offset by an opposing force, the Earth's atmospheric temperature would now be far too hot for life to be thriving. Yet, life on Earth is currently positively thriving.

Life on Earth is currently positively thriving because it has responded with an opposing force, in order to keep its atmospheric temperature favourable for its continued flourishing. This opposing force is the pulling of carbon dioxide out of the Earth's atmosphere and its storage under the Earth's surface. This is why atmospheric carbon dioxide concentrations have fallen from 5000ppm to only a few hundred ppm.

When the human species started to have a significant impact on the Earth, at the start of the Industrial Revolution, life on Earth had offset increasing incoming solar radiation to such an extent that the level of atmospheric carbon dioxide had fallen from 5000ppm to 280ppm. That is a colossal reduction! Through this colossal reduction life on Earth has been able to prevent its atmospheric temperature from rising to a level that is too hot for life to survive and thrive.

What does this mean for the future of life on Earth? The amount of solar radiation reaching the Earth is only going to keep on increasing in the future. How can life on Earth continue to thrive in the face of this continued continuous increasing? What has happened in the past clearly cannot continue. In the past, a fall from 5000ppm to 280ppm has been required to offset the increasing solar radiation and thereby maintain the Earth's atmospheric temperature. Atmospheric carbon dioxide cannot fall much lower than this. One can easily comprehend that, if what has been needed in the past, was to continue to work in the future, then atmospheric carbon dioxide concentrations would not only need to fall to zero, they would ultimately have to fall to minus several thousand!! That is obviously not possible!

Atmospheric carbon dioxide concentrations cannot fall much below 280ppm because the thriving of life on Earth requires a significant amount of carbon dioxide to be in the Earth's atmosphere. The existence of life requires plants to perform photosynthesis (take carbon dioxide from the atmosphere) and to respire (put carbon dioxide into the atmosphere), and the existence of animals entails putting carbon dioxide into the atmosphere via respiration. So, obviously, a thriving life-bearing planet, a planet which has trillions upon trillions of flourishing life-forms, will need to have quite a lot of carbon dioxide in its atmosphere. Following a reduction to 280ppm, from 5000ppm, there isn't much scope for atmospheric carbon dioxide levels to fall much further. If life is to continue to survive and thrive then another solution, another opposing force, is desperately needed.

This is the cause of the main threat to life on Earth. Increasing solar radiation has almost overwhelmed the ability of life on Earth to maintain its atmospheric temperature, to keep it down to a level that is conducive for its continued thriving, through reducing its 'greenhouse effect' via atmospheric carbon dioxide removal.

Life on Earth is in grave peril; it needs to be saved. It needs to be saved from increasing levels of incoming solar radiation.

How can life on Earth be saved?

Life on Earth is currently in grave danger. For hundreds of millions of years life on Earth has successfully maintained the conditions that it requires for its continued flourishing through offsetting increasing solar radiation by reducing atmospheric carbon dioxide concentrations. We can call such an offsetting/reducing life on Earth's 'first line of defence'.

The fall in atmospheric carbon dioxide concentrations to 280ppm means that the 'first line of defence' is speeding towards eternal impotence. If life on Earth is to continue to survive, and thrive, it requires a new way to maintain its atmospheric temperature in the face of forever increasing levels of incoming solar radiation.

There is only one way in which life on Earth can be saved. Life on Earth needs to develop the capability to deflect incoming solar radiation, thereby preventing it from reaching the Earth's atmosphere. Such deflection is life on Earth's 'main line of defence'.

The human species

The human species is that part of life on Earth which has the ability to bring forth the 'main line of defence'. In other words, the human species is the saviour of life on Earth. In the absence of the human species there would be no 'main line of defence'. And, without the possibility of a 'main line of defence' life on Earth would be lurching towards oblivion. If the human species did not exist, the day of obliteration for life on Earth would be nigh.

Thankfully, life on Earth has complexified to the point that is the human species, and it thus has the ability to bring forth the 'main line of defence'. Life on Earth, thanks to the human species, has a long and glorious future.

The 'guilt-tinged' age of the technological explosion

Given that the human species is the saviour of life on Earth, why is it that we are currently faced with a pandemic of urgings to 'save the planet' from the destructive and destabilising activities of the human species?

Bringing forth its 'main line of defence' requires life on a planet to go through a technological explosion. More specifically, a stage in Solar-Systic unfoldment is reached when, due to the inevitable impending impotence of its 'first line of defence', the epoch of the technological explosion is needed if a thriving life-bearing planet is to continue to flourish; without it, life on that planet is doomed. However, it is also an atrocious epoch; a truly ghastly period of Solar-Systic unfoldment. The bringing forth of technology entails the bringing forth of immense disruption, large destabilisations and tremendous suffering. These things come into being as the bringer forth of technology explores its planet, as it learns how to master the resources that are at its disposal, as it hones its abilities as the master modifier of its Solar System.

As technology explodes on a planet, suffering also explodes on that planet. The master modifier itself suffers immensely, as do the life-forms which the master modifier inevitably sees as resources for its use. The nature of the master modifier, the human species, is to master modify. So, the master modifier master modifies.

However, a point eventually comes, in the midst of the technological explosion, when the master modifier realises that its master modifications have had consequences that it did not foresee. The master modifier was just acting in accordance with its nature; it didn't think about consequences; indeed, until very late in the day there were no significant consequences to be thought about. Almost in the blink of an eye the human population size exploded, technology exploded, and the master modifier suddenly had a massive disruptive and destabilising impact on the Earth. When the master modifier comes to realise that its past actions have serious consequences, when it comes face-to-face with the suffering that it has brought into being, when it comes to realise the extent to which it has destabilised the biogeochemical cycles of the Earth, then it naturally becomes overwhelmed with guilt.

There is fear, there is anguish, and there is despair. What have we done to the planet? Why are we so selfish, and greedy, and stupid? We need to stop what we are doing! We need to rein in our activities!! Such is the initial response of the master modifier when it comes to realise the extent of its impact on the Earth in the 'guilt-tinged' age of the technological explosion.

This initial response, this guilt, explains the urgings of those who want us to 'save the planet' from the destructive and destabilising activities of the human species. Those who urge us to rein in our activity in order to 'save the planet' are overwhelmed by the guilt that comes to pervade human culture in the 'guilt-tinged' age of the technological explosion.

The partial reversal of the 'first line of defence'

In the epoch of the technological explosion the master modifier rampantly utilises all of the resources that it can fruitfully make use of on its host planet. These resources include the fruits of the 'first line of defence'. In other words, the master modifier releases the fossil fuels that were previously stored underground as life on Earth successfully regulated its atmospheric temperature in the pre-human era through the pulling of carbon dioxide out of its atmosphere. The fruits of the 'first line of defence', what we call 'fossil fuels', power the technological explosion which brings forth the 'main line of defence'.

The release of the fossil fuels that were previously stored under the surface of the Earth is a partial reversal in life on Earth's 'first line of defence', as it brings forth its 'main line of defence' (the fall in atmospheric carbon dioxide concentrations from 5000ppm to 280ppm, gets partially reversed as it increases to over 400ppm). In other words, the increasing level of atmospheric carbon dioxide that is occurring at the moment, due to the human release of previously stored fossil fuels, is a sign that the master modifier is in full swing; it is a sign that life on Earth is well on its way to bringing forth its 'main line of defence'.

If one was to observe the pattern of atmospheric carbon dioxide concentrations that have occurred on the Earth, a massive fall over a prolonged period (from 5000ppm to 280ppm on the Earth), followed by an abrupt spike (from 280ppm to over 400ppm on the Earth), on any life-bearing planet, then this would be a sign that all is well on that planet. It indicates that the 'first line of defence' has done its job and that life on that planet is well on its way to bringing forth its 'main line of defence'.

If one is observing the atmospheric carbon dioxide concentrations on a distantly located life-bearing planet, and there is a prolonged fall over a long period of time, but there is no sign of an abrupt spike, then one should be deeply and increasingly worried for the life on that planet. One would be praying to see such a spike soon, for without it one would know that the life that resides there would be hurtling towards extinction. In other words, one would be praying to see global warming that has been initiated by that planet's master modifier.

Environmentalism

We have seen that the realisation of the magnitude of the human impact on the Earth gives rise to a sense of guilt concerning what the human species has done. When this sense of guilt has arisen it inevitably leads to the phenomenon which has been dubbed 'environmentalism'. In the face of anger, guilt and despair, emotions are aroused, and growing numbers of humans seek to alleviate their guilt through reducing their impact on the Earth. They also seek to urge other people to do the same; and, those who do not act in accordance with these urgings become the 'enemy'. Such urgings, and the associated creation of the 'enemy', is perfectly understandable and inevitable. Such urgings play a part in bringing about a truly sustainable and peaceful world.

However, as we have seen, if life on Earth is to survive, and thrive, it requires its master modifier, the human species, to bring forth its 'main line of defence'. And, this bringing forth means that the disruptive epoch of the technological explosion needs to be passed through. So, whilst a pervasive sense of guilt emerges in the epoch of the technological explosion, the human species hasn't done anything that it should feel guilty about. The human species has been acting the way that it has in order to bring forth the 'main line of defence' so that it can save life on Earth. Saving life on Earth is a noble endeavour; it isn't something to feel guilty about!!

Of course, when it is fully enmeshed in the 'guilt-tinged' age of the technological explosion, the overwhelming majority of the human species does not realise that its actions are contributing to the outcome of saving life on Earth, so all that is left is the guilt! Feeling such guilt is therefore natural and normal; after all, in the ghastly epoch that is the age of the technological explosion, the master modifier of its planet and solar system inflicts a tremendous amount of suffering and death onto its fellow planetary life-forms.

Human suffering

The airplane crashes. Hundreds of people die. Suffering. Death. Mourning. Misery for friends and family.

The school bus gets ploughed into by a lorry. All the children die. Suffering. Death. Mourning. Misery for friends and family.

A machine gun massacre. The nuclear bomb explodes. The roller coaster comes off its rails. The bungee jump cord snaps. The explosive device detonates. The car ploughs into the pedestrians. The air of the city is full of fumes and smog created by technological society; the people who breathe the air become immensely sick.

The masses flock to live in concrete jungles and spend their time immersed in concrete and engulfed in technology; they spend most of their time staring at the screens of technological devices. Is this good for their health? Of course it isn't! They need forests, flowers, fresh air and the sea; yet, the masses flock to the concrete jungles and stare at the screens of technological devices! Such a lifestyle brings them immense mental ill-health and physical ill-health. Many cannot take it anymore; they throw themselves off the tall buildings humans have built; they shoot themselves with the guns humans have created; they shoot others with the guns humans have created; they cut their wrists with sharp knives and razor blades; they drink alcohol until they get into a state of oblivion.

The masses eat highly processed, chemical-filled foods which have been produced by mass agriculture, and which have been deeply degraded by being zapped in microwave ovens and decimated in deep fat fryers. The masses eat animals, which their bodies were not designed to eat. The masses make themselves terribly ill through all of this. Why don't they eat fresh, organic, healthy food – fruits, vegetables and grains? Why do they eat degraded food that degrades their mental and physical health?

Why do humans live like this? Why do humans do all of this to themselves? If humans had a simple life; for example, if they transported themselves under their own steam, like other animals, using their own two feet, there would be no carnage. Currently over a million humans a year are killed in road accidents; slaughtered by motor vehicles on roads. The precious children who were in the bus that was ploughed into by the lorry would still be alive if humans transported themselves by foot!

It is not human nature to transport itself by foot. It is not human nature to live a simple technology-free life. The human is not just another animal. The human species is the master modifier of its planet, and solar system; it has no option but to act in accordance with its nature and to master modify. Transform! Modify! Create! Analyse! Investigate! Utilise! Modify some more! Invent! Invent better and faster! Modify more and more and more!!!

In the atrocious epoch that is the technological explosion the human species is bringing forth increasingly complex technologies, for the benefit of life on Earth, so that it can fulfil its cosmic purpose, but in so doing it suffers tremendously. The human species is currently in the stage of Solar-Systic unfoldment where it is unleashing technology, but it has yet to master the technologies that it is bringing forth. The technologies are running amok; consequently, humans are suffering and dying in their hordes. Millions upon millions of humans are being slaughtered due to the technologies that humans have brought into being!

Such is the atrocious epoch that is the age of the technological explosion. It is the extreme magnitude of the technology-induced misery, suffering and death, both for the human species, and for the rest of life on Earth, that makes this epoch so atrocious.

The future will be better. The human species will eventually both master its technologies and master itself so that it doesn't abuse the technologies that it has brought forth. The suffering of the human species will eventually be brought down to a bearable level as our Solar System continues to unfold and we move out of the atrocious epoch that is the age of the technological explosion.

The enlightened humans of the future

If we fast forward a few thousand years into the future, we can appreciate that the human species will be much more enlightened, immensely more spiritually developed. The humans alive at this time will, no doubt, look back at the epoch of the technological explosion and think to themselves: "How could our ancestors possibly have used and abused their fellow planetary life-forms to such an extent? How could they possibly have lived with themselves? Causing such abuse must have pushed them into insanity! They must have been overcome with guilt concerning what they were doing!"

Yet, these more enlightened humans of the future will understand perfectly well why this atrocious epoch needed to be passed through; for, without it, they know that the peaceful, sustainable planet that they inhabit would never have come into existence. They know that without the 'main line of defence' life on Earth would have been obliterated. They know that if, in the epoch of the technological explosion, the human species wasn't deluded concerning its place in our Solar System, if it didn't see its fellow life-forms as inferior resources to be used, if it didn't see itself as the selfish destroyer rather than as the precious saviour, if it didn't become overwhelmed with guilt, then the master modifier would not have been able to bring forth the wonderful 'main line of defence'.

One cannot master modify, exploit, enslave, slaughter, factory farm, vivisect, eat, that which one sees as similar to oneself; the master modifier, in the epoch of the technological explosion, has to see itself as special, as superior; it needs to believe that its fellow planetary life-forms are very different to itself, when in reality they are not.

Delusion

In the age of the technological explosion the human species becomes deluded. There are three strands to this delusion. Firstly, the human species is deluded because it does not realise that it is the precious saviour of life on Earth; rather, it believes that it is fundamentally a destructive influence on the Earth, as far as life is concerned. Secondly, the human species is deluded into believing that the non-human animals it shares the planet with are fundamentally different to humans, when in reality they are not. Thirdly, the human species is deluded into believing that it has a fairly good understanding of the nature of the Universe, the world that surrounds it, when in reality it does not. Our immediate concern is to explore the first of these three strands; we will return to the other two strands in due course.

The technological explosion brings forth immense suffering for both human and non-human life-forms, it entails colossal planetary transformation which negatively impacts both human and non-human life-forms, and it entails serious destabilisation of planetary biogeochemical cycles which threatens the continued flourishing of life. In the midst of these phenomena it is exceptionally hard for the human species to see that it is actually the precious saviour of life on Earth.

In order to escape this delusion one needs to come to appreciate that life on Earth's 'first line of defence' is close to being overwhelmed, and that the future thriving of life on Earth requires it to be supplanted with its 'main line of defence'. Furthermore, one needs to appreciate that the bringer forth of the 'main line of defence' is life on Earth's precious saviour, the master modifier that is the human species.

Having recognised this theoretical potential escape route it is important to realise that this delusion cannot easily be escaped from within the epoch of the technological explosion. For, within the epoch of the technological explosion the human species needs to be deluded concerning its cosmic purpose, because without such a delusion it wouldn't be able to fulfil its cosmic purpose. If the human species was not deluded, if humans *en masse* could see things as they truly are, then they could not possibly bring themselves to be the initiators of the immense suffering and large destabilisations that are necessary parts of the age of the technological explosion.

Our relationship to the non-human life-forms of the Earth

Let us now focus on the second strand of the state of human delusion. This is the very widespread belief that humans are fundamentally different to the non-human life-forms of the Earth. In order to power the technological explosion the human species has needed to see everything in its surroundings, including its fellow planetary life-forms, as resources for its use. The advancement in human knowledge, and the advancement in human technological prowess, has been immensely bolstered by the use of non-human life-forms in a plethora of ways. As human culture has powered forward these life-forms have been dissected, observed, vivisected, ridden into war, put to work in agriculture, and put to the sword as a major food source to power the activities of an escalating human population.

The human species has had to justify to itself this mass exploitation of its fellow planetary life-forms. In other words, the human species has had to believe that there is a chasm between itself and its fellow planetary life-forms; it has had to tell itself that these life-forms are inferior, that they are lacking important attributes that humans possess. You will be familiar with many variants of this attempted justification. They are just animals:

They cannot think.

They are not conscious.

They lack self-awareness.

They don't have emotions.

They don't have culture.

They cannot feel pain.

They don't have a mind.

They lack language.

They cannot rationalise.

They lack feelings and are just machines.

They lack any moral worth.

They cannot use tools.

They lack creativity.

They lack intelligence.

They cannot suffer.

This attempted justification for the existence of a chasm between the human species and all of its fellow planetary life-forms is clearly both immensely widespread and totally absurd. A plethora of non-human life-forms on the Earth have all of the attributes listed above. There is no 'superiority-making attribute'; there is no attribute that a human possesses which all non-human planetary life-forms lack, the possession of which makes the human fundamentally different to, and superior to, the non-human. It is just that, in the epoch of the technological explosion, the human species has had to believe that such an attribute exists in order to justify its exploitation of these life-forms, and thereby maintain its sanity in the face of the immense suffering that it is inflicting upon them.

The only thing that meaningfully distinguishes the human species from the rest of the life-forms of the Earth is that the human species occupies a special place in the unfolding of life on Earth. The human species is the master modifier of its planet which has the cosmic purpose of bringing forth the 'main line of defence' in order to save life on Earth. So, the human species is superior to all of the other life-forms of the Earth, but not because of the possession of any particular attribute which particular humans possess and particular non-humans lack.

Of course, one could, quite rightly, say that the ability to bring into being the 'main line of defence' is a superiority-making attribute that the human species has, which the non-human species of the Earth lack. However, one needs to keep in mind that such an ability doesn't represent a fundamental chasm in the attributes possessed by individual humans on the one hand and individual non-human life-forms on the other hand, of the sort which has been hypothesised by the human species in order to justify the mass exploitation of its fellow planetary life-forms. The various justifications for the human utilisation of non-human animals, the hypothesised chasm between the human and the non-human, is simply a necessary delusion in the epoch of the technological explosion.

If humans were to realise the extent of the intelligence, awareness, suffering and compassion that exists in the life-forms that they have been exploiting, then they couldn't possibly bring themselves to engage in such exploitative acts. Whilst the human species needs to be deluded in the epoch of the technological explosion, it will come to fully appreciate the true capabilities and attributes of its fellow planetary life-forms when our Solar System has unfolded to the point that is 'the day of the realisation of cosmic purpose'.

Our cosmic purpose

The purpose of the human species is to bring forth life on Earth's 'main line of defence'. We can refer to the 'main line of defence' as 'the sufficient deployment of technology' because when technology has complexified to the point that enables it to be deployed to create the 'main line of defence' then the future of life on Earth will be secure.

The human species is currently working tirelessly, whether it realises it or not, to bring forth the wonderful technological fruit that is 'the sufficient deployment of technology'. Humans are forever bringing forth increasingly complex technologies, forever learning more and more about the nature of the Earth and the constituents that make up the Universe, forever utilising technologies, forever fulfilling their potential and pursuing their passions; through all of these things human culture is irresistibly powered forward.

This tireless work will ultimately lead to the bringing forth of 'the sufficient deployment of technology'. There are other technological fruits that the human species is bringing forth to protect life on Earth, such as the ability to deflect massive asteroids which are on a collision course with the Earth. However, whilst such fruits are immensely tasty side-dishes, the prize fruit, the main course, is the desperately needed 'main line of defence' – 'the sufficient deployment of technology'.

If the human species was lazy, if the vast majority of humans spent their days lounging around and not working, not learning, not enquiring, not seeking to fulfil their potential and better themselves, then life on Earth would be doomed. However, such laziness would be a ghastly denial of our nature. We simply couldn't be that lazy; such laziness is not us! We are the master modifier of our Solar System. Our nature is to master modify. We cannot not do this. We have to work tirelessly until 'the sufficient deployment of technology' has been brought into being and life on Earth has thereby been saved.

'The day of the realisation of cosmic purpose'

In the future there will come a time when knowledge of the nature of human cosmic purpose will pervade the human species. When 'the sufficient deployment of technology' has come into being, when the human species has fulfilled its cosmic purpose, it will then be possible for the human species to come face-to-face with its purpose.

On 'the day of the realisation of cosmic purpose' the nature of human cosmic purpose, the reason why humans have been working so tirelessly for so long to enable the bringing forth of increasingly complex technologies, becomes crystal clear for all to see.

On 'the day of the realisation of cosmic purpose' the human species joyously celebrates its achievements in saving life on Earth through bringing forth 'the sufficient deployment of technology'. On this day the human species comes to fully appreciate the unfolding journey of our Solar System as it has striven to bring forth life and then to maintain life. This is a wonderful journey that involves the gallant efforts of the 'first line of defence' and its inevitable supplantation by the 'main line of defence' that is magnificently brought into being through the tireless work of the human species.

Human-induced global warming

In the midst of the epoch of the technological explosion, human-induced global warming is inevitably seen as a wholly bad thing; this is because there is no widespread realisation that it is an extremely welcome partial reversal in life on Earth's 'first line of defence' as it brings forth its desperately needed 'main line of defence'.

Human-induced global warming is not a wholly good thing; it can lead to undesirable outcomes, to death and suffering, in particular locations. These undesirable outcomes are part of the price that needs to be paid for the bringing forth of the 'main line of defence'. These undesirable outcomes arise due to the inevitable delusion of the human species. If the entire human species was instantaneously released from its state of delusion, then our cosmic purpose would be speedily accomplished, and global warming and its associated undesirable outcomes would be averted.

Minimising suffering

If we were to fully accept our nature, as master modifiers, we would then realise that reining in our activities, having a minimal impact on the Earth, is not possible in the epoch of the technological explosion.

If we were to fully accept the nature of our cosmic purpose, to bring forth 'the sufficient deployment of technology', if we could see that this deployment was badly needed, and required for the future thriving, and ultimately the future survival, of life on Earth, then we could push forward at speed and bring forth 'the sufficient deployment of technology' as soon as possible.

If we were to accept our nature, and acknowledge our cosmic purpose, then we could minimise the suffering of human and non-human life-forms in the future by positively embracing our destiny. We could positively embrace the challenge of bringing forth 'the sufficient deployment of technology', and we could seek to bring it into being with the utmost haste.

The bringing into being of 'the sufficient deployment of technology' means that there will be no global warming, there will be no increase in the Earth's atmospheric temperature, so there will be no suffering to human and non-human life-forms resulting from global warming.

The longer that we are deluded, the longer that we deny our nature, the longer that we resist our destiny, the longer that we take to fulfil our cosmic purpose, the greater will be the amount of global warming that occurs, and, consequently, the greater will be the amount of avoidable suffering and death that emanates from global warming.

The folly of focusing on fossil fuel emission reduction efforts

Those who want us to deny our nature, those who are blinded to our cosmic purpose, believe that the phenomenon of global warming can be addressed through reducing, or ceasing, our use of fossil fuels. Consequently, they urge us to immediately significantly reduce, and/or to speedily eliminate, fossil fuel use.

Why do so many people believe this? Why do they urge us to take this course of action? The reason is that these people believe that global warming is primarily a single factor scientific issue. What does this mean? Scientists have realised that increased greenhouse gas concentrations in the Earth's atmosphere lead to an enhanced 'greenhouse effect' and cause global warming. To take a single factor scientific perspective is to jump directly from this realisation to the conclusion that the appropriate human response, the way to stop global warming, is to focus on fossil fuel emission reduction efforts, to focus on stopping putting greenhouse gases into the Earth's atmosphere.

We are slowly coming to appreciate the folly of this simplistic approach. We can see that life on Earth needs us to bring forth its 'main line of defence'. It is our nature to act in such a way that enables this to happen. Acting in accordance with our nature leads to the technological explosion, which entails a partial reversal in life on Earth's 'first line of defence', through the release of the previously stored fossil fuels, and leads ultimately to the fulfilment of our cosmic purpose – the bringing into being of life on Earth's 'main line of defence'.

We need to come to fully appreciate why the appropriate human response to stop global warming is to bring forth the 'main line of defence' as soon as possible. Furthermore, we need to understand why any attempt to eschew the 'main line of defence' is a harmful waste of time and energy which will increase the suffering of both human and non-human life-forms.

As we are currently in the midst of the 'guilt-tinged' age of the technological explosion, it is understandable that there is a widespread belief that we should be focused on fossil fuel emission reduction efforts. However, such efforts are a great distraction which could lead to the obliteration of life on Earth. We don't have time to waste, time to be distracted, life is in peril and we need to bring forth 'the sufficient deployment of technology' as a matter of urgency.

The global warming situation that life on Earth currently faces is a potent combination of two forces: non-human-induced global warming (the incoming solar radiation which has almost overwhelmed the 'first line of defence') and human-induced global warming (the release of the fruits of the 'first line of defence'). The combination of forces, factors and time-lag effects currently in play relating to these two phenomena means that we need to fulfil our cosmic purpose with extreme haste.

Focusing on fossil fuel emission reduction efforts is an attempt to renounce our nature; it is an attempt to eschew our cosmic purpose. The longer that the human species focuses on such efforts, rather than focusing on bringing into being 'the sufficient deployment of technology', the greater is the probability that life on Earth will be obliterated. Such is the folly of focusing on fossil fuel emission reduction efforts.

Human-induced global warming is not a single factor scientific issue

It is of immense importance to fully appreciate that human-induced global warming is not a single factor scientific issue. It is all too easy to believe that it is. It is true that scientists have realised that increased greenhouse gas concentrations in the Earth's atmosphere lead to an enhanced 'greenhouse effect' and thereby are a force for human-induced global warming. However, this realisation does not directly lead to an indubitable conclusion considering the appropriate human response. It is not the case that one can jump straight to the conclusion that the appropriate human response is to focus on reducing fossil fuel emissions.

There are many other factors in play which determine the appropriate human response. Some of these factors are scientific in nature, other of these factors are non-scientific in nature. Yet, in reality the scientific and the non-scientific meld together in intricate ways, because a simple scientific finding can be interpreted in diverse ways when it is placed in broader contexts.

Two types of global warming

It is of the utmost importance to acknowledge, when considering the appropriate human response to the phenomenon of human-induced global warming, that there are actually two types of global warming. When one has acknowledged that there are two different types of global warming – the force for non-human-induced global warming (increasing solar radiation) and the force for human-induced global warming (the release of the fruits of the 'first line of defence') – then one immediately has a broader context within which to consider the appropriate human response.

Such an acknowledgement means that one can appreciate that life on Earth has responded to non-human-induced global warming through its 'first line of defence' which has created what we call fossil fuels. One can appreciate that the 'first line of defence' cannot continue to save life on Earth because the reduction in atmospheric carbon dioxide concentrations from 5000ppm to 280ppm means that its power, its effectiveness, has ebbed away. The 'first line of defence' is fast approaching eternal impotence in terms of its ability to offset forever increasing levels of incoming solar radiation. The next step is to appreciate that life on Earth needed to complexify to the point that is the human species, and the technological explosion, if it was to continue to survive and thrive. When one appreciates this, then one can see that the human release of the fruits of the 'first line of defence' – fossil fuels – is a partial reversal in life on Earth's 'first line of defence' as it brings forth its 'main line of defence'. One can also appreciate that the 'main line of defence' is 'the sufficient deployment of technology' – technology which deflects incoming solar radiation, thereby preventing it from reaching the Earth's atmosphere. Finally, one can come to see that such a bringing forth is the cosmic purpose of the human species.

It is not helpful to try and conceptually disentangle human-induced global warming from non-human-induced global warming. Indeed, it is dangerous to believe that human-induced global warming is a single factor scientific issue that can be adequately addressed through fossil fuel emission reduction efforts, whereas non-human-induced global warming is a separate issue which operates over a vastly longer timescale. Timescales are irrelevant. What is relevant, what is important, is to appreciate the forces that exist in the present moment; the question of how long it took for these forces to come into existence is not important.

The force for non-human-induced global warming is exceptionally powerful at the present moment, and it is interweaving with the force for human-induced global warming to create an immensely potent force. This immense force could very easily lead to the obliteration of life on Earth if the human species does not embrace its nature and its cosmic purpose. We need to power forward and proactively take control of the planet, not cower away and attempt a sheepish retreat through attempting to reduce our fossil fuel emissions.

One can appreciate that the conclusion that we have just reached is an inevitable melding of the scientific and the non-scientific. The knowledge that we have concerning the existence of non-human-induced global warming is scientific knowledge. The knowledge that we have concerning the existence of human-induced global warming is scientific knowledge. However, by themselves these two pieces of knowledge are not very fruitful. These pieces of knowledge need to be weaved together into a broad context in order to reveal the way that our Solar System, and the Earth, is unfolding. When one can see this context, then one can clearly see the nature of human cosmic purpose and how this relates to the phenomenon of global warming.

The 'non-pent-up force' for human-induced global warming

In order to appreciate the situation that we are currently facing one needs to clearly see that there are two forces for human-induced global warming – the 'non-pent-up force' which we will consider here, and the 'pent-up force' which we will consider shortly.

The 'non-pent-up-force' has two elements. The first element arises due to the fact that when the human species moves the fruits of the 'first line of defence' from underground storage to the Earth's biosphere, some of these released fossil fuels move to a new home in the Earth's atmosphere. In other words, there is a partial reversal in life on Earth's 'first line of defence' as atmospheric carbon dioxide levels to start to nudge upwards. This nudging upwards is a 'non-pent-up force' for human-induced global warming.

The second element arises due to the way that the human species, in the epoch of the technological explosion, has fundamentally modified the biosphere of the Earth in ways that make it harder for life on Earth to keep its atmospheric temperature down. In other words, it represents the ways that the activities of the human species have further weakened the effectiveness of life on Earth's rapidly deteriorating 'first line of defence'.

In terms of this second element, we are primarily talking about deforestation. We are talking about the intentional transformation of wonderful precious forests, which are a key part of the 'first line of defence', to habitats which suit the human species, such as cities / concrete jungles and barren expanses that are farmed by mechanised agriculture. Forests are a precious conduit between atmospheric carbon dioxide drawdown and fossil fuel creation. As far as the 'first line of defence' is concerned, concrete jungles and agricultural barren expanses are a total waste of space!

In the atrocious epoch that is the age of the technological explosion the human population size explodes, and this exploding population increasingly masters the art of intricately transforming the biosphere of the Earth to suit its needs, through the creation of such things as chainsaws, combine harvesters, trains, skyscrapers, airplanes, cars, motorways, slaughter houses and the mass utilisation of non-human life-forms as agricultural commodities. This mass transformation of the composition of the Earth's biosphere means that when the fruits of the 'first line of defence' are released into the biosphere, that atmospheric carbon dioxide concentrations become higher than would otherwise have been the case. In other words, the human transformation of the Earth's biosphere is part of the 'non-pent-up force' for human-induced global warming.

All of this means that in the epoch of the technological explosion a massive 'non-pent-up force' for human-induced global warming is created through the two elements that we have considered – the movement of fossil fuels from underground storage to their new home in the Earth's atmosphere, which is exacerbated by the wholesale human transformation of the Earth's biosphere from lush forests to lifeless concrete and barren expanses that are farmed by industrial agriculture.

It is crucially important to ruminate on the fact that in the epoch of the technological explosion, as life on Earth brings forth its 'main line of defence', which it needs to replace its rapidly deteriorating 'first line of defence', that the 'first line of defence' becomes even weaker. This deterioration is due to the second element that we have been considering: the biospheric transformation initiated by the human species. This increasing weakness means that we are currently living in a period of immense danger. We need to further appreciate exactly why this is so; we need to come to appreciate all of the forces that are contributing to this immense danger; we need to behold the bigger picture, the overall situation, which encompasses our place in the unfolding of our planet, and the unfolding of our Solar System. In short, we need to urgently move past the extremely simplistic view of human-induced global warming as a single factor scientific issue.

The 'pent-up force' for human-induced global warming

We are coming to realise that human-induced global warming is not a single factor scientific issue. We need to deepen this realisation by considering the existence of the 'pent-up force' for human-induced global warming. This force has already been unleashed but it has yet to reveal its global warming impacts.

When a master modifier, in the epoch of the technological explosion, releases the fruits of its planet's 'first line of defence' – fossil fuels – these fruits do not make a wholesale move into a new home in the atmosphere of that planet; if they did they would cause very significant immediate global warming. Rather, what happens is that an immense force for future global warming is created, and, worryingly, it is easy to be blinded to the existence of this force. We need to fully appreciate that this time-lag 'pent-up force' for future global warming exists. If we are blinded to it, it could be curtains for the future of life on Earth.

When fossil fuels are burned about half of the released carbon dioxide gets sucked straight into the planet's ocean where it sets off on a lengthy journey through what, on the Earth, we have called *the thermohaline circulation*. The thermohaline circulation involves dense cold water sinking at high latitudes and travelling very slowly through the ocean depths until it eventually reaches the northern Indian Ocean and the northern Pacific Ocean where it resurfaces. After the carbon dioxide sinks into the thermohaline it takes from 100 years to 1000 years to emerge into the atmosphere. This means that almost all of the carbon dioxide that has become temporarily stored in the thermohaline circulation since the start of the Industrial Revolution has yet to emerge into the atmosphere. Oh no!!! When the bulk of this carbon dioxide starts to be released it is set to turn from a 'pent-up force' for human-induced global warming to a bringer of a large increase in the temperature of the Earth's atmosphere.

Worryingly, the world to which this carbon dioxide is returning is very different to the world that it left. It departed from a world full of lush rainforests, which had a cool ocean and a vigorous thermohaline circulation. It will be released into a human-transformed biosphere in which rainforests have been replaced with concrete jungles and agricultural barren expanses; a human-transformed biosphere which has a warmer ocean, and a weakened thermohaline circulation. This vastly changed world is a place in which carbon dioxide accumulates in the atmosphere. For, the rainforest 'carbon sinks' have been removed. And the global warming that has occurred since the start of the Industrial Revolution has both warmed the oceans, making them less able to absorb carbon dioxide, and significantly weakened the thermohaline circulation through melting icebergs and glaciers, which also means that far less carbon dioxide can be absorbed into the oceans. So, when the long-travelling oceanic carbon dioxide starts to gush out of the thermohaline circulation it will start accumulating in the atmosphere of this vastly different, human-transformed world. This transformed biosphere massively exacerbates the magnitude of the rise in atmospheric temperature that is set to occur. So, the human transformation of the biosphere, in the age of the technological explosion, is part of both the 'non-pent-up force' for human-induced global warming and the 'pent-up force' for human-induced global warming.

If this large increase in atmospheric temperature was to occur, it is likely to trigger other large-scale discontinuities; for example, a runaway 'greenhouse effect' resulting from the destabilisation of methane clathrate reservoirs. By around the year 3000 – when all of the carbon dioxide that is currently stored in the thermohaline has been released – the atmospheric temperature of the Earth could very easily be way too hot for the existence of the human species, and for any other complex life-forms.

Furthermore, if this large increase in atmospheric temperature was to occur, life on Earth would be forever doomed because its 'first line of defence' would have been fatally impaled, and its attempt to attain its 'main line of defence' in time would have failed. The Earth's biosphere would have transitioned into a new hotter state and the future would only ever involve increasingly hot atmospheric temperatures due to the forever increasing strength of the force for non-human-induced global warming. The Earth would never again be able to support the existence of complex life-forms. Of course, as we are considering a 'pent-up force', this would be the case even if humans were to stop using fossil fuels today.

Let us repeat the point, because it is a crucial one. We are considering the future impacts of a force that has already been created, a 'pent-up force', so it matters not a jot whether or not humans reduce their future fossil fuel emissions, or even eliminate them completely. Such efforts are futile; they are just a great distraction. If such efforts distract us sufficiently they will be the cause of the obliteration of life on Earth!!! What a travesty that would be.

Acknowledging the existence of the 'pent-up force' for human-induced global warming is essential. The existence of this force means that fossil fuel emission reduction efforts are not a solution to the human-induced global warming situation that we are currently facing, let alone the overall global warming situation that we are currently facing. This overall situation is a combined force resulting from the interfusion of the force for non-human-induced global warming with the two forces for human-induced global warming – the 'non-pent-up force' and the 'pent-up force'. A fruitful expression of the way that these forces combine to create the overall global warming situation that we are currently facing can be found in the section: *Two Forces for Global Warming*.

We can increasingly appreciate that human-induced global warming is not a single factor scientific issue.

We can increasingly appreciate the folly of focusing on fossil fuel emission reduction efforts.

We need to immediately get our heads out of the sand and start to act in an appropriate way in order to save life on Earth! Let us utilise the talents and abilities that our Solar System has endowed to us and bring forth 'the sufficient deployment of technology' with extreme haste!

The 'marginal difference tipping point fallacy'

You might find it hard to accept that the human endeavour to reduce fossil fuel emissions is futile. I have been making the case that such efforts are feeble, and are totally incapable of dealing with the magnitude of the forces for global warming that are lined up against us. Nevertheless, you might be thinking to yourself that this endeavour has to be a good thing, because any tiny marginal difference could determine whether or not an atmospheric tipping point is passed. According to this way of thinking, it is possible that the lifestyle of a single human could make the difference. If this particular human had a high carbon footprint a tipping point could be passed, causing the Earth's atmospheric temperature to shoot upwards; whereas, if they had a low carbon footprint, the tipping point would not be passed. There is a sound logic to this way of thinking. Atmospheric tipping points do exist, and marginal differences can cause a tipping point threshold to be passed.

So, why is the 'marginal difference tipping point' phenomenon not applicable to our current situation: the forces for global warming that are lined up against us, and our emissions of fossil fuels? The reason is that we are not in a position in which we need to be concerned with a marginal difference causing a tipping point to be passed. This is because our current position is one in which tipping points are set to be overshot by a ginormous margin. This means that we need a correspondingly ginormous response. We shouldn't be concerning ourselves with tiny marginal differences.

Imagine two different scenarios. These scenarios involve a human, a small stone and water. In our first scenario, a human is submerged in water and is drowning. Thankfully they find a small stone to stand on and this enables them to raise their head above the water thereby enabling them to breath. In this scenario, the marginal difference that is the stone has made all the difference. In our second scenario, our human is standing on a beach as a tsunami is heading towards them. They rationalise that in order to survive they need to move to higher ground. This is definitely a sound rationalisation. There is a tipping point here; this is the height of the ground at which if our human is at or above it the tsunami will not hit them, but if they are below this height the tsunami will smash into them. Our human does move to higher ground; they step onto a small stone that is on the beach. Not surprisingly, they drown. In this scenario the 'marginal difference tipping point' phenomenon is not applicable. It is a fallacy. Stepping onto the stone is a total waste of time; it is a feeble response that is totally incapable of dealing with the situation. This second situation is the situation that we are in when it comes to our current fossil fuel emissions – these emissions are the stone, whilst the forces for global warming that are lined up against us is the tsunami.

The 'marginal difference tipping point fallacy' is a natural bedfellow of existing in the 'guilt-tinged' age of the technological explosion. People who feel guilty naturally want to feel like they can do something to make a difference: *"If I get a bicycle rather than a car I might save the planet"*. Really! Really!!

The time for action has arrived. We need to be powering forward. We need to be proactively taking control of the Earth's atmospheric temperature. We shouldn't be cowering away and sheepishly attempting to repudiate our nature and our cosmic purpose. We shouldn't be faffing around expending our energies on fossil fuel emission reduction efforts.

The timing of the bringing forth of the master modifier

I have been painting a picture of our unfolding Solar System. This picture involves the human species, the master modifier, the saviour of life on Earth, the bringer into being of the 'main line of defence', coming into existence at the exact same time as the Earth's 'first line of defence' is frail and deteriorating, and close to collapse. Is this timing a happy coincidence?

Obviously, if our Solar System unfolded to the point at which the 'first line of defence' collapsed before the master modifier came into being, then life on Earth would be decimated; the Earth's atmospheric temperature would forever be too hot for the thriving of life.

Alternatively, it is possible to imagine that at the point of Solar-Systic unfoldment that the master modifier came into being that the 'first line of defence' was still robust, in rude health and in its relative youth. In this case, there would be no urgency to bring forth the 'main line of defence'.

It is obviously in the interests of life on a planet to bring into being its master modifier as soon as it possibly can, because this significantly boosts its survival chances. Yet, this feat is no easy task. When life first evolved on the Earth it was extremely simple in its composition, which means that it had a very limited ability to intricately mould its surroundings. Life on Earth has gone through a process of complexification from this extremely simple composition to the exceptionally complex arrangement that is the master modifier. The master modifier is able to mould its surroundings in the most amazingly delicate and intricate ways. The master modifier is the endpoint of an exceptionally long journey which has involved life on Earth ever so gradually increasing its ability to modify its surroundings in increasingly delicate and intricate ways. So, whilst the ideal scenario for life on Earth is to bring forth its master modifier as soon as life arose on the planet, this was obviously an endeavour that was going to take an exceptionally long period of time.

Whilst it takes life on a planet an exceptionally long stretch of time to bring into being its master modifier, the 'first line of defence' of life on a planet also lasts for an exceptionally long stretch of time. This means that it is exceptionally easy to comprehend why there is likely to be a similarity in timing between the bringing into being of life on Earth's master modifier and the weakening and deterioration of life on Earth's 'first line of defence'. This similarity in timing is certainly not a cause of great surprise. Perhaps this similarity in timing reveals a deeper order at work in our Solar System; perhaps things are brought into being when they are needed, and things work out for the best.

What does this mean for our current situation? We can come to fully appreciate that we have inevitably been bought forth into existence at the time that life on Earth's 'first line of defence' is extremely frail and badly deteriorating. We can accept that our purpose is to bring forth 'the sufficient deployment of technology' to save life on Earth and give it a long and glorious future. We can realise that achieving our cosmic purpose involves us utilising the abilities that our Solar System has provided us with; such abilities have already produced the technological explosion. We can come to see that this involves us causing immense disruption to the Earth, including releasing the fossil fuels which are the stored fruits resulting from the functioning of life on Earth's 'first line of defence'. We can come to appreciate that our focus of concern should not be human-induced global warming. Our focus of concern should be the overall situation that we face, given all of the above.

Précis

We have been coming to appreciate that human-induced global warming is not a single factor scientific issue – one cannot jump directly from the scientific appreciation that humans are enhancing the 'greenhouse effect' through the release of fossil fuels, to the conclusion that the appropriate response is to reduce and eliminate fossil fuel use. Rather, the phenomenon of human-induced global warming needs to be seen against the backdrop of the unfolding of our Solar System, as one factor which is intertwining with many other factors and forces to give rise to the overall situation that we currently face.

We have made good progress in our exploration of this overall situation, this broader context within which the phenomenon of human-induced global warming is situated, by considering three intertwining forces – the force for non-human-induced global warming, the 'non-pent-up force' for human-induced global warming, and the 'pent-up force' for human-induced global warming. We now need to consider another extremely powerful force that is in play – 'the force that is human nature'.

'The force that is human nature'

In order to more fully appreciate the situation that we are currently in, and consequently the course of action that we need to take, we need to reflect on our own nature.

We are not alien to our fellow planetary life-forms, alien to our planet, alien to our Solar System. It is not as if we travelled from an alternate Universe and plonked ourselves on the Earth. Rather, we are a fundamental expression of the nature of the Universe. The Universe, our Solar System, our planet, our fellow life-forms, they all brought us into being; they evolved us. We are an outgrowth of all that is. Our abilities, our talents, our skills, our feelings, are all fundamental expressions of the underlying nature of the Universe, our Solar System, our planet, and the life which brought us into being.

Furthermore, we didn't evolve as a fluke. Every life-form, and every part of the Universe, modifies its surroundings to the best of its ability, using the talents/skills that are at its disposal. This obviously applies to us. Indeed, we are the zenith of the ability to modify; we are the master modifier of our Solar System. Our skills, our abilities, our talents, and our passions, are gifts that have been endowed to us by our Solar System, and it is our nature to utilise these things as best we can. We didn't create these things, let us repeat, they are gifts from our Solar System. If we use these gifts, that cannot be a bad thing; we weren't given these gifts to cause mischief! We were brought into being in order to master modify; we were brought into being so that we could make use of the gifts which we have been provided with.

The natural/normal state of affairs is for a human to seek to make use of their abilities, skills, talents and passions. Some humans are gifted scientists/engineers/inventors. Some humans are gifted entrepreneurs / business people / politicians. Some humans are gifted actors. Some humans are gifted teachers/writers/academics. Some humans are gifted footballers/athletes. Some humans are gifted illustrators/mathematicians/musicians. Some humans are gifted nurses/doctors/surgeons. Some humans are gifted spiritual leaders / comedians / artists. Some humans are gifted at fossil fuel extraction. Some humans are gifted environmentalists. We need to be utilising all of our gifts. It is, of course, possible for a particular human to have a range of abilities/skills/talents/passions that spans all of the above domains; however, this is not normal. The overwhelming majority of humans are gifted at some things and terrible at other things. Furthermore, some humans are immensely passionate about things which other humans are wholly indifferent to. It is good that these differences exist. More than this, these differences are wonderful! The human species has been provided with the diverse range of abilities/skills/talents/passions that it needs in order to fulfil its cosmic purpose and bring forth 'the sufficient deployment of technology'.

Why do humans act the way that they do? Humans seek to be happy; they seek to have a pleasant and meaningful life. How is this accomplished? A human needs to know what they are passionate about. A human also needs to know what their particular abilities/skills/talents are. It is highly likely that there will be a significant crossover between these two things. This means that there will be lots of ways in which a human can deploy their abilities/skills/talents in accordance with their passions. If such a deployment occurs then we can say that a human is fulfilling their individual potential. To talk of human nature is to talk of the plethora of human passions and their deployment through particular abilities/skills/talents. In other words, to talk of human nature is to talk of humans fulfilling their potential.

The activity of individual humans as they seek to fulfil their potential is in itself a force. When we consider the activities of all the humans that exist we can start to get a grasp of how this force operates at the planetary level, across the Earth. The cumulative day-to-day activity of every human across the planet generates 'the force that is human nature'.

In the age of the technological explosion 'the force that is human nature' is dominated by the master modification activities of the human species. In this age, the natural state for the overwhelming majority of humans is to act so as to fulfil their potential through engaging in activities which entail immense modifications of their surroundings. This could be through driving their car, partaking in airplane flights, using a combine harvester, playing the piano, constructing a skyscraper, operating a chainsaw, mowing their lawn, talking on their mobile phone, buying/consuming/hoarding an immense plethora of things, and bringing forth or utilising a whole range of technological devices. This colossal force, this unquenchable thirst for master modification, is unstoppable in the epoch of the technological explosion.

The unstoppable nature of this force in the epoch of the technological explosion is something to be celebrated, not something to be feared. For sure, it leads to planetary transformation, it involves mass deforestation, it entails the release of stored fossil fuels, it triggers the destabilisation of biogeochemical cycles, and it generates immense death and suffering for both human and non-human life-forms. However, this unquenchable thirst, this colossal force, this exuberant expression of the fundamental nature of the Universe, ultimately leads to the bringing into being of life on Earth's desperately needed 'main line of defence' – 'the sufficient deployment of technology'. In the absence of this force, life on Earth would be doomed. Whilst the epoch of the technological explosion is brutal for the particular life-forms that exist within it, it is simultaneously an immensely precious epoch for life on Earth.

'The force that is human nature' needs to be acknowledged, accepted and embraced. It represents our fundamental nature, it motivates us, it flows through us, and its continued unleashing is required if life on Earth is to have a long and glorious future. There is no point seeking to repudiate this force, or wish it away; such an endeavour will do much more harm than good. The master modifier has to master modify!

The concoction of forces that inevitably threaten life on Earth at precisely this moment

Our focus of concern needs to be the delicate interplay of forces that are currently in play. There are a perilous concoction of forces that are inevitably currently combining at precisely this moment in the evolution of our planet and Solar System. We are living at the moment of extreme danger for life on Earth.

The force for non-human-induced global warming is currently exceptionally powerful as it has almost overwhelmed life on Earth's 'first line of defence'. The extreme potency of this force is inevitably bolstered, in deeply worrying ways, by the activities of the master modifier as it has brought forth life on Earth's technological explosion. The activities of the master modifier have further weakened life on Earth's already extremely frail 'first line of defence' – its ability to regulate its atmospheric temperature – through the transformation of wonderful forests into desolate concrete jungles and barren expanses that are farmed by mechanised agriculture. Furthermore, the master modifier has released the fruits of the 'first line of defence' and has thereby unleashed a potent, and potentially devastating, 'pent-up force' for future human-induced global warming, which will be unleashed into the Earth's atmosphere when the slow-moving carbon dioxide starts to gush out of the thermohaline circulation, in the coming decades and centuries, into a vastly changed world. This is a human-transformed world in which carbon dioxide accumulates in the atmosphere, rather than the land/oceans. So, the Earth's atmospheric temperature is set to shoot up. This means that there will be an immense threat to life on Earth when the slow-moving oceanic carbon dioxide reaches its atmospheric destination.

The forces that we have just considered form an exceptionally potent concoction; they pose an immense threat to the continued flourishing of life on Earth. This concoction is about to get even more potent! We need to add to it 'the force that is human nature'. As we are in the midst of the age of the technological explosion 'the force that is human nature' is inevitably propelling us down the path of increasing future master modification as humans continue to fulfil their potential. This means that the immediate future entails a greater human impact on the planet: greater perturbations, greater destabilisations, greater transformations, and greater danger in the present moment.

We need to acknowledge this perilous concoction of forces. We need to embrace our destiny. Life on Earth needs us to embrace our cosmic purpose. The time of danger is now. The time for action is now. Will we succeed or will life on Earth perish?

Tinkering with the 'greenhouse effect'

Most people are unaware of the concoction of extremely dangerous forces that are currently threatening the continued flourishing of life on Earth. Most people see global warming as a single factor scientific issue and hence believe that a 'small scale' response, or a series of such responses, is an appropriate response to the situation that we currently face. For example:

Let's plant lots of trees!

Let's try and cut our fossil fuel emissions!

Let's try and reduce the number of airplane flights!

Let's encourage people to buy electric cars!

Let's sell our shares in fossil fuel companies!

Let's invest in renewable energy!

Let's try and recycle more and buy fewer things!

Let's try and persuade the government to decarbonise the economy!

If you become aware of the perilous concoction of forces that currently exist, then you will realise that all of this is totally hopeless and tragically woefully inadequate. These 'small scale' responses are really a type of 'return to the past' thinking. This goes something like: *we have messed things up through our destabilising activities, so we now need to try and rein our activities in, to 'leave things to nature', so that we can restore the balance that existed in the past.* What we desperately need is a 'big scale' response; we need to embrace our nature as the master modifier of our planet; we need to power forward and take control; we need to increase our presence on the planet, not rein it in. If we fail to do this we will, through our inaction, be cruelly condemning, and ultimately extinguishing, the life that has arisen on the Earth.

All 'small scale' responses involve attempting to tinker with the 'greenhouse effect'. Such tinkerings don't go anywhere close to the scale of the response that is needed to deal with the concoction of forces that are aligned against us at this moment in time. As you are well aware, the reduction in atmospheric carbon dioxide concentrations to 280ppm at the start of the Industrial Revolution means that the 'first line of defence' has almost worn out. On top of this we have inevitably unleashed both the 'non-pent-up force' for human-induced global warming, and the 'pent-up force' for human-induced global warming, so the situation is now extremely dire.

We urgently need to stop thinking about tinkering with the 'greenhouse effect' and instead focus on bringing forth 'the sufficient deployment of technology' as soon as possible. However, whilst we are bringing forth 'the sufficient deployment of technology' we might need to deploy our master modification abilities to buy a little more time not through a tinkering, but through a major manipulation of the 'greenhouse effect' which entails initiating a massive drawdown of carbon dioxide from the atmosphere. This might entail the creation of ocean seaweed forests. However, such a massive drawdown would be nothing more than a less than ideal temporary patching up of the fatally impaled 'first line of defence'. It is not a long-term solution and it could have unwelcome side-effects; yet, it might be needed if we cannot bring 'the sufficient deployment of technology' into existence in time. What is really needed is not any kind of manipulation of the 'greenhouse effect'; what is needed is the 'main line of defence' itself – 'the sufficient deployment of technology'.

Tinkering with the 'greenhouse effect' will not give life on Earth a long and glorious future; it won't even give life on Earth a short and glorious future. What life on Earth urgently needs is its most powerful weapon – 'the sufficient deployment of technology'. Let's get to work!

The next 300 years

Given that the amount of fossil fuels burned by humans has been enormously high in the past 100 years, and was also extremely high in the preceding 100 years, and given that roughly half of the carbon dioxide thereby released was sucked into the thermohaline circulation, where it embarks on an oceanic journey ranging from 100 years to 1000 years, after which it emerges into the atmosphere of the vastly changed world, you can quickly do the maths and work out that the next few hundred years are going to be interesting!

The next 300 years are the time of real danger for life on Earth. This is the time when the force for human-induced global warming is at the zenith of its potency. This is mainly due to the 'pent-up force' coming to fruition, but in this period the 'non-pent-up force' also reaches its peak strength as the epoch of the technological explosion reaches its zenith. And, of course, the force for non-human-induced global warming is already exceptionally potent, having almost overwhelmed life on Earth's 'first line of defence'. *The combination of these extremely powerful forces, this perilous concoction, is something that we should be terrified about.*

We are the glorious saviours of life on Earth, but in order to fulfil our cosmic purpose we have had to be deluded; we have had to bring forth the brutal epoch of the technological explosion; we have had to bring forth immense suffering both for ourselves and for our fellow planetary life-forms; we have had to perturb the Earth's biogeochemical cycles in a way that exacerbates the possibility that life on Earth will be decimated in the near future. This has all been necessary so that we can bring forth 'the sufficient deployment of technology' in order to save life on Earth, giving it a long and glorious future.

We need to immediately get our collective heads out of the sand; we need to start acting in an appropriate way in order to save life on Earth! Let us free ourselves of the erroneous belief that the situation that we currently face can be resolved through fossil fuel emission reduction efforts. We are living at the pivotal moment in the unfolding of life on Earth, if we continue to focus on fossil fuel emission reduction efforts, and ignore 'the sufficient deployment of technology', then life on Earth will be decimated in the near future. There is no time to waste. Let us fulfil our destiny. Let us fulfil our cosmic purpose. Let us bring forth 'the sufficient deployment of technology' with the utmost haste.

'The sufficient deployment of technology'

We have seen that if life on Earth is to survive and have a long and glorious future then it needs its frail and deteriorating 'first line of defence' to be replaced by its 'main line of defence'. In other words, it needs its master modifier, the human species, to bring forth 'the sufficient deployment of technology'. The 'main line of defence' entails the deployment of technology to deflect incoming solar radiation, thereby preventing it from reaching the Earth's atmosphere. It is our cosmic purpose to bring forth 'the sufficient deployment of technology' and the sooner that we do this the better. My role is to highlight the need for 'the sufficient deployment of technology' and to explain why such a deployment is not only a necessity, but also an exceptionally precious and wonderful thing; it is not something to be afraid of, or something to be resisted.

What, exactly, does 'the sufficient deployment of technology' look like? What does it physically entail? These are questions for the engineers, for the experts in technology; they are not questions for me; my objective is simply to make clear the need for 'the sufficient deployment of technology', so that the engineers and policymakers can be encouraged to push full steam ahead and bring it into being.

Any technology that can deflect incoming solar radiation to the Earth in a reliable and controlled way, thereby actively regulating the Earth's atmospheric temperature to keep it down to a level that is favourable for the continued thriving of life, could be 'the sufficient deployment of technology'.

Let us briefly consider a few possible physical instantiations of 'the sufficient deployment of technology'. A space sunshade is the most obvious candidate. The sunshade would seemingly need to be placed at the L1 Lagrangian point between the Earth and the Sun, which is located 1.5 million kilometres above the Earth. At Lagrangian points a smaller object will maintain its position in relation to two larger bodies that are in orbit. The sunshade at the L1 point could either be a single large object or a plethora of small objects.

In terms of the former possibility, in 1989 James Early designed a 2000 kilometre glass occulter lens which could be placed at the L1 point. This lens would disperse solar radiation away from the Earth's atmosphere. In a similar vein, in 2004 Gregory Benford calculated that a concave rotating Fresnel lens placed at the L1 point, which was 1000 kilometres across, and a few millimetres thick, would deflect up to 1% of incoming solar radiation to the Earth's atmosphere. Rather than deploying a lens, in 1997 Edward Teller, Lowell Wood and Roderick Hyde proposed that a 3000 ton thin wire mesh diffraction grating could be placed at the L1 point.

In terms of the latter possibility, in 2006 Roger Angel's design involved placing 16 trillion small disks at the L1 point. This design involves each disk having a 0.6 metre diameter and a thickness of 5 micrometres. This number and design of disks would deflect 2% of the incoming solar radiation to the Earth.

If we zoom forward in time, if we envision the state of our Solar System in a few thousand years, we can be confident that there will be technological devices at the L1 point which deflect solar radiation away from the Earth's atmosphere. In the immediate future, while we are still escalating our technological expertise, there is also the possibility for using technology to prevent solar radiation from reaching the lower atmosphere of the Earth (the troposphere) through manipulating the composition of the Earth's upper atmosphere (the stratosphere). For example, aircraft could be used to put sulfate aerosols into the stratosphere where they would reflect incoming solar radiation back into space.

Could such a manipulation of the stratosphere be a long-term solution? Could it be 'the sufficient deployment of technology'? It is very hard to imagine that this could be the case. Such a manipulation is likely to have severe, and undesirable, side-effects. It would also be hard to attain the required levels of control and reliability. Furthermore, it is a relatively short-term manipulation which requires regular replenishment. The safest and the most elegant form for 'the sufficient deployment of technology' to take is the deployment of a technological sunshade at the L1 point.

Embracing extreme uncertainty

It is now time for us to return to our consideration of the state of human delusion which exists in the age of the technological explosion. We earlier explored the first two strands of this state of delusion. The third strand of this state of delusion is that the human species believes that it has a better understanding of the nature of the Universe, and a better understanding of the unfolding of the biogeochemical cycles of the Earth, than it actually has. We will consider in great detail the nature of the Universe, and this third strand of human delusion, in this section: *Human Purpose & the Universal Pursuit of Ecstasy.* For now, we can note that one consequence of this 'understanding deficit' is that the human species consistently underestimates the level of uncertainty that exists. Sometimes the level of this underestimation is ginormous!

In the realm of environmentalism, this results in a range of forecasts which the human species believes to be very diverse and broad in scope, but which are all actually in a very narrow and conservative range. For example, the IPCC has produced a range of forecasts concerning the possible level of the Earth's atmospheric temperature by the year 2100. This range of forecasts is formulated within a belief system which necessitates that the human species believes that it has more knowledge than it actually has. What this means is that even the most pessimistic forecasts could be overshot, and not just by a small margin, but by an enormous margin!

If we could fully comprehend the extent of the uncertainty that exists then this would be most advantageous. Embracing extreme uncertainty leads to the realisation that there is an extreme need for a vastly increased ability to control and modify. Such an increased ability to control and modify provides the means through which a high level of future uncertainty can be adequately dealt with by the human species for the benefit of life on Earth. In other words, embracing extreme uncertainty paves the way for the speedy bringing forth of both 'the sufficient deployment of technology' and other technological protectors of life on Earth.

Cherishing the living

We might be the master modifier, the zenith of the evolutionary progression of life on Earth, the saviour of life on Earth, the most precious life-form in our Solar System, but we don't have to see our fellow planetary life-forms as inferior, as resources to be used and eaten. We needed to see our fellow planetary life-forms this way in the past, we needed to use and abuse non-human animals in the past, but there is certainly no need to do so now. When an individual life-form dies this is a cause for sadness; when it survives this is a cause for celebration. Let us cherish each and every one of these immensely precious life-forms.

Paradise on Earth

When we have passed through the painful era that is the epoch of the technological explosion, when we have fulfilled our cosmic purpose and brought forth 'the sufficient deployment of technology', then the Earth will transition into a new phase of its existence.

Life on Earth has a long and glorious future. The super-advanced technology that will be maintaining the atmospheric temperature in order to keep it favourable for the continued thriving of life on Earth enables this. In this new phase of Solar-Systic unfoldment every human will appreciate the nature of human cosmic purpose because 'the day of the realisation of cosmic purpose' occurs. In this new era the life-forms on the Earth will be living together more harmoniously than they ever did in the previous stages of the Earth's history.

In this new phase, the age of the spiritual explosion, the human species is living sustainably and it has the utmost respect for all of the life-forms that it shares the Earth with. Paradise will have come into existence on the Earth.

Celebration

Wouldn't it be nice if we could celebrate our existence? We are the saviours of life on Earth! The survival of our non-human brothers/sisters is our responsibility; we need to take care of them through bringing forth 'the sufficient deployment of technology'. Oh, if we could see all this, we could then celebrate our existence! Oh, if we could realise how important and special we are! Oh, if we could see the cosmic responsibility that rests on our shoulders! Could we possibly bear to come face-to-face with this immense responsibility?

Our Solar System is celebrating us; it is joyous at our arrival. Yet, we are miserable! We are in a state of despair! We inevitably tell ourselves that we are selfish and greedy and that our presence on the Earth is destructive and harmful. We tell ourselves that we are the enemy of our non-human brothers/sisters. How deceived we are. How wonderfully deceived!

One day humans will be celebrating. One day will be 'the day of the realisation of cosmic purpose'. On this day, every human will appreciate the value of all humans, all cultures, all perspectives and all life-forms. On this day, every human will see the nature of human cosmic purpose; they will see the wonderful journey of Solar-Systic unfoldment laid out before them in its entirety.

Human Purpose & the Universal Pursuit of Ecstasy

In the section *Human Purpose & the Universal Pursuit of Ecstasy* we will explore in greater depth some of the themes that we have explored here, whilst also placing these themes in a much larger context by exploring the very nature of the Universe and the way that it unfolds through time. This exploration penetrates through to the heart of the third strand of human delusion, a strand which we have barely been able to get below the surface of here.

In *Human Purpose & the Universal Pursuit of Ecstasy* I explain why the term 'human species' is most appropriately used not as a biological term; rather, the term should be used to refer to the master modifier of its host planet and solar system. The human species = the master modifier. This is why human-induced global warming is an inevitable stage of unfoldment of *any* thriving life-bearing planet. As you are well aware, human-induced global warming is a sign that all is well on a planet; it is a sign that life on that planet is well on its way to bringing forth its 'main line of defence' which it needs to replace its inevitably deteriorating 'first line of defence'.

It is worth noting that there might appear to be a slightly different tone in the two sections. In *Human Purpose & the Universal Pursuit of Ecstasy* I take a fatalistic approach to the way that human culture, the Earth, and our Solar System, unfolds. I say that we are cosmic puppets who are powerless to meaningfully change the way that things will unfold in the future. Whereas, in this section, I am making an urgent 'call for action', a plea to radically change our approach to the phenomenon of global warming, so that we can change course and thereby minimise the amount of suffering and death that occurs, whilst also giving life on Earth a long and glorious future. Such a plea, you might think, only makes sense if we are able to change the way that things will unfold in the future in a significant way. This difference in tone is not contradictory. For, the very writing of this book, and any consequences that might flow from this event, are part of the fatalistic unfolding of our Solar System!

In this section I have sought to get straight to the point, straight to the heart of the issue of how the human species relates to the other life-forms on the Earth, and why this means that we are the saviour of life on Earth which has a cosmic purpose to urgently fulfil. I have also sought to clearly explain why the overwhelming majority of humans alive at the moment inevitably see the human species in a negative light, due to their existing in the midst of the disruptive epoch that is the 'guilt-tinged' age of the technological explosion. If you would like a deeper perspective on the nature of the Universe, and the way that our Solar System unfolds through time so as to bring forth the human species, then you will enjoy *Human Purpose & the Universal Pursuit of Ecstasy.*

Embracing our destiny / Embracing your destiny

I hope that you have come to appreciate that the future of the human species, and the future of life on Earth, will be determined by the way that we respond to the potent concoction of forces that inevitably threaten life on Earth at precisely this moment.

If we continue to be deluded, if we continue to be blinded to our cosmic purpose, if we continue to believe that we are the destroyers of life on Earth, rather than its precious saviour, then the future is extremely bleak. It is the current predominance of this belief, a fleeting predominance which is an inevitable aspect of the 'guilt-tinged' age of the technological explosion, which has led to the pandemic of urgings to 'save the planet' from the destructive and destabilising activities of the human species. In turn, this has resulted in the widespread belief that the appropriate human response to the phenomenon of human-induced global warming is to attempt to rein in our activity, to reduce our fossil fuel emissions, to reduce our impact on the Earth and to 'leave things to nature'. Such a response, as we have seen, has devastating consequences. It results in needless global warming, needless climate change caused by global warming, needless immense suffering, and needless death.

Let us escape our delusion! Let us wake up to the fact that human-induced global warming is not a single factor scientific issue! Let us come to appreciate the wider context in which the phenomenon of human-induced global warming is situated! Let us realise our nature! Let us embrace our destiny! Let us fulfil our cosmic purpose! Let us create a future in which there is no more needless suffering and death emanating from our delusion!

We need to increase our presence on the Earth; we need to power forward; we need to extend the reach of our master modification abilities. We need to move full steam ahead and embrace the challenge of deploying advanced human-created technologies to regulate the temperature of the Earth's atmosphere. If we attempt to rein in our presence on the Earth, to 'leave things to nature', then the day of decimation for life on Earth will be nigh. If we were to rein in our presence on the Earth, we would be tragically letting down the life that has arisen on the Earth. What an utter disappointment we would be!

Life on Earth needs us!

Life on Earth needs you!

Go forth; activate your passions; utilise your abilities, skills and talents; fulfil your potential!

Go forth; throw off the shackles of your delusion!

Go forth; embrace the glorious future in which the human species saves life on Earth through the speedy bringing into existence of 'the sufficient deployment of technology'!

Human Purpose

& the

Universal Pursuit of Ecstasy

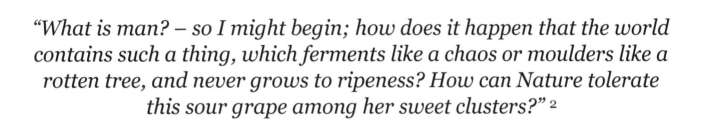

"What is man? – so I might begin; how does it happen that the world contains such a thing, which ferments like a chaos or moulders like a rotten tree, and never grows to ripeness? How can Nature tolerate this sour grape among her sweet clusters?" [2]

Friedrich Hölderlin

[2]Hölderlin, F. (1797) ' Hyperion' in Eric L. Santner (ed.), *Hyperion and Selected Poems,* New York: Continuum, 1990, p. 35.

Dear Friedrich

The human species is ecstasy and its coming into being is where it ends; for, ecstasy is the endpoint of a universal pursuit. That which troubles you, the chaos and the rottenness, enables the bringing into being of wondrous life-giving fruits. The need for these joyous delicacies explains the tolerance of the sourness that precedes them. Rest assured my friend, the human species will reach ripeness, but be patient, the universal pursuit of ecstasy cannot be rushed!

The human species = ecstasy

What is the human species?

What is ecstasy?

What is the purpose of the human species?

Our journey

You are about to embark on a journey, a journey into the depths of the Universe. What follows is an account of the most fundamental issues relating to human existence. As our journey progresses you will gain insights into the nature of reality, what it means to be human, the meaning of life and how to best live your life. You will come to see how the human species relates to the non-human life-forms that we share the planet with. You will come to appreciate why the human species is the most precious life-form that has ever, and will ever, exist in our Solar System. You will come to understand why the human species has inevitably dominated the planet and why it has rampantly utilised our fellow life-forms as resources. You will come to appreciate the unfolding journey of our Solar System, and of human culture, as inevitable war and exploitative domination gradually come to be replaced by universal peace and respect for all life-forms. You will come to see why the 'environmental crisis' and human-induced global warming / climate change are not bad events; rather, they are signs that life on Earth, and in our Solar System, is positively thriving. You will also come to appreciate the vastly differing stories that we, as a species, tell ourselves about our place on the Earth. Some of these stories are optimistic and positive in tone, whilst others are pessimistic and present a gloomy view of the human species.

One such gloomy story that is currently quite popular is the view that the human species is an unwelcome guest on the Earth, some kind of destroying parasite, and that the Earth would be far better off if the human species became extinct. In this story the human species is conceptualised as an entity that brings immense suffering to the non-human life-forms of the Earth for its own selfish benefit. As our journey progresses you will come to appreciate both why this story has been created and why it is totally misguided. You will eventually come to see that the human species is the most precious form of life in our Solar System, and that the purpose of the human species, the reason that it came into existence, is intimately connected to the phenomenon of global warming. You will come to see that rather than being the bringer of suffering to non-human Earthly life-forms for its own selfish benefit, the fundamental underlying truth is that the human species suffers vastly more than any of the other life-forms on the Earth, so that it can be the saviour of non-human Earthly life.

Let's elucidate this a little. Our realisation will be that if the human species didn't initiate any global warming of our atmospheric temperature then this would mean that we had failed in our cosmic mission; we would have failed to fulfil our purpose, and thus life in our Solar System would be forever doomed. This would be a tragic outcome. The phenomenon of human-induced global warming is not a bad event, some kind of tragedy, a terrible injustice to non-human life, as is commonly portrayed. In order to appreciate why this is so, one needs to be aware of the bigger picture within which the phenomenon is situated. In other words, one needs to appreciate the full swathe of the evolutionary unfolding of our Solar System, of our planet and of human culture. Another way of putting this is to say that one needs to come face-to-face with the universal pursuit of ecstasy. Our journey, a journey that has just begun, is a journey of understanding into the nature of the universal pursuit of ecstasy. For, all parts of the Universe are continuously engaged in the universal pursuit of ecstasy. If one solely sees the phenomenon of human-induced global warming, and our appropriate response to it, solely through an extremely narrow scientific lens, in total ignorance of the universal pursuit of ecstasy, then one will be hopelessly lost and deluded. It won't be simply that one's head will be buried in the sand; rather, one's entire body will be submerged in the depths of the sandy beach.

Words are a tragic substitute for insight into the nature of the Universe, yet they are also a wonderful tool to provide understanding if they are appropriately used. Humans are rational, questioning creatures, so it is likely that you will have some doubts concerning the truth of that which follows. 'Why should I believe this?' you might think to yourself. Try, for now, to put these doubts, this questioning approach, aside. You might want to go further, you might want to try and forget everything that you have ever been taught/read/learnt and start afresh. The closer you are to such a state, the better you will be able to come to know the universal pursuit of ecstasy. This is because you are likely to encounter ideas which clash, sometimes violently, with your present understanding. When this clash occurs a natural desire to stand by your long-standing ingrained beliefs is likely to arise thereby resulting in you being blinded to the truth of that which follows. Our objective is to radically change the way that you see the world around you and this cannot be achieved in a few short paragraphs. We need to embark on a journey of progressive insight, a journey which involves gathering a multitude of pieces of understanding which will ultimately fuse together. As we reach our wonderful destination these diverse pieces will majestically come together to form a magnificent jigsaw, a wonderful vision of the unfolding Universe and an appreciation of the extremely important place of the human species within this unfoldment.

Our journey is long and winding, and bumpy in places, but if you see it through to its exquisite destination then it could be the trip of a lifetime. Let us start where we need to begin. We need to consider the Universe that appears to us when our sensory organs are engaged, and we need to appreciate that the Universe itself is very different to this appearance.

The Universe that appears to us

The Universe is not as it appears. Nothing is as it appears. The things that we take to constitute our Universe, things such as planets, trees, dogs, humans, cars and carrots, are not real distinct objects. These things constitute the Universe that appears to us when we actively construct our surroundings. We construct a Universe of individual objects, a Universe of distinct things; it inevitably follows that these appearances tell us very little about the Universe itself. The things that appear to us are clearly part of the Universe, but we need to appreciate the difference between appearance and the underlying nature of things.

To start with, we need to consider the difference between a static appearance and the way that this part of the Universe changes through time. A static appearance does not provide any knowledge concerning the nature of a particular part of the Universe. However, some knowledge of a particular part of the Universe that appears to us (we can call such a part a 'thing') can be attained through acquaintance. There are two routes to acquaintance – observation over time and interaction. All things become acquainted with other things to varying degrees, because all things interact with other things. So, as a human is a thing, this means that every human becomes acquainted with various things to various degrees. Some things are highly acquainted with some humans, whilst being an utter mystery to other humans.

Appearance versus knowledge via acquaintance

Take two things, a human that you have known intimately for a very long time and the planet Venus. It is fairly safe to assume that you have more knowledge of the human than you have of the planet. Why is this? It is not because you are also a human! It is because you have spent much more time observing the human than you have spent observing the planet, combined with the fact that you have a much higher level of interaction with the human than the planet. To observe a thing is to monitor it over time, chiefly via visual perception. To interact with a thing is to be in its vicinity; the closer the proximity between two things, the greater will be their level of interaction. The higher level of acquaintance between you and the human that you know intimately means that you have more knowledge of this human than you have of the planet Venus.

If one was to be presented with two static images, an image of a thing which appears to be a human and an image of a thing which appears to be the planet Venus, then one could easily assume that one has more knowledge of the 'human' than of the 'planet'; it is exceptionally easy to mistakenly conflate the appearance of a thing with knowledge of a thing. Yet, what does one really know about this 'human'? One really has no knowledge; one just has a static appearance, an image. Whilst the image appears to be an image of an actual living thing, a human, it could actually be an image of a waxwork from Madame Tussauds.

Imagine that in a distant galaxy an extremely intelligent entity exists that is very different in appearance from the human bodily form. If this entity was presented with the two images, the 'human' and the 'planet Venus', what would they make of these appearances? This entity would have no knowledge of either thing, because they lack acquaintance with either thing. More than this, they would have no assumption that they have more knowledge about one thing than they do the other. Both things would be utterly mysterious to them. One cannot acquire knowledge of a thing from a static appearance.

Such a conclusion is familiar enough in the everyday human realm. One does not get to know another human through looking at a static appearance. One gets some knowledge of another human via acquaintance through observing them from moment to moment. The greater the observations, the greater the acquaintance.

Observation over time and interaction penetrate appearance and delve into the nature of a thing. This applies to all things, to humans, to birds, to bonobos, to helicopters, to snowdrops and to planets. Observation over time and interaction penetrate appearance and generate knowledge of the nature of things. No thing is isolated; every thing interacts with other things; so, every thing has knowledge of the nature of other things.

Human prejudice

How much knowledge of things can be attained as a result of observation over time? In the human realm the answer to this question is tied up with the phenomenon of human prejudice. Humans typically have pre-existing expectations which generate particular beliefs concerning the nature of the things that they observe; this is what we mean by human prejudice. These pre-existing expectations naturally tend to separate humans, to isolate humans, to differentiate the nature of the human from the nature of the non-human. We will only be in a position to fully appreciate the cause of this natural tendency when we have progressed a long way through our journey. For now, we can say that in our current epoch the human sees itself as fundamentally different, as fundamentally superior, to all of the other things that it observes in the Universe, and that this belief inevitably leads to a type of human prejudice which fundamentally distinguishes humans from the rest of the Universe.

The existence of human prejudice means that observation over time is tainted, rather than pure. The alternative to observing things through the prism of human prejudice is to simply observe things as they are, to be with them as they are, wholly free of expectation. Human prejudice generates a lot of false beliefs. It can very easily cause a human to assume that things that resemble them have certain qualities, whereas things that do not resemble them do not have these qualities. For example, another human resembles one, so one assumes that this thing has Quality X, whereas the planet Venus and a carrot do not resemble one, so one assumes that these things do not have Quality X. In this way human prejudice generates a vast array of beliefs concerning the nature of things that are false. To talk of human prejudice is, effectively, to talk of such false beliefs. However, some of the pre-existing expectations that arise from human prejudice will accord with reality.

Human prejudice is immensely widespread. What are the implications of this? It means that knowledge of the nature of things is severely restricted because the overwhelming majority of humans have flawed expectations/beliefs concerning the nature of most of the things that they observe. These flawed expectations/beliefs act as a barrier which prevents knowledge of things being obtained; they keep the believer in the domain of the prison of mere appearance.

In the face of human prejudice, to what extent can things be known through observation over time? Even when a human is pervaded by human prejudice they can still gain some knowledge of things by observing them over time. This can be a casual process of observing a thing, or it can be a more detailed and systematic process of scientific analysis. However, it goes without saying that if one is under the influence of human prejudice that one's knowledge of things will always be severely limited. It matters not a jot how many years of observations one racks up, one will never get remotely close to knowing things as they are. One will, for sure, get to know things better, but getting to know things better is very different from knowing things as they are. We can here utilize an oft-used metaphor. One could spend one's entire life getting to know an iceberg increasingly better, yet on one's deathbed one could have still only seen the tip of the iceberg. To know things as they are is to see the entire iceberg. Is it possible to see the icy masses that lie below the ocean surface, and thereby get to know things as they are? In order to answer this question we need to consider the way that the visual sense presents a surrounding world.

The limits of the visual sense

The question that we are spending some time addressing is: Can things be known as they are? This is, of course, a question that has vexed philosophers ever since philosophers came into existence. The question is so vexing because the things that appear to humans are inevitably presented via the human visual sense and this inevitable presentation has two implications.

The first implication is that a different visual sensory system would present different things. This is partly a matter of the scale/size of the objects presented. Some visual sensory systems present a very fine-grained view of things, with an immense number of things existing in a particular place. Whereas, other visual sensory systems are more coarse-grained, and present much fewer things in the same place. To mentally grasp this we can simply ponder, and contrast, the immense variety of eyes that exist throughout the animal kingdom. Alternatively, you could imagine looking through a microscope at some things and then consider how all things would appear to you if your visual sensory system was set up so as to permanently/continuously view things in this fine-grained fashion. The inevitable observation of things from a singular-grained perspective, through a particular visual sensory system, itself feeds the phenomenon of human prejudice.

The second implication is that humans can only observe the exterior of things; they lack the ability to observe the interior of things. So, one might observe another human who is clutching their leg whilst screaming and howling in apparent agony, but one cannot observe any painful qualitative feeling states in this human. In order to attempt to get to know this thing, this human, better, one automatically moves from the realm of the observation of the exterior to the realm of judgements/beliefs concerning the interior. So, one's past experiences of oneself, and one's past experiences of observing and interacting with other humans, is likely to lead one to believe that the interior of this human currently contains very unpleasant qualitative feelings. Another way of putting this is to say that the belief arises from a very limited level of knowledge concerning the nature of things which is strongly bolstered and reinforced by human prejudice. It is obvious that this belief is a mental judgement, an episode of calculated guesswork, arising from limited knowledge concerning the nature of things due to the limits of the visual sense.

If one was observing the same clutching/screaming/howling whilst watching an actor in a soap opera, then one would surely make a very different judgement and have a very different belief; one would believe that the human does not actually contain painful qualitative feelings

despite the almost identical observations. Given this, it is very easy to appreciate how one can end up with a flawed belief concerning the nature of things. Can one be confident that one is observing a thing that contains painful qualitative feelings, rather than a thing that is pretending to do so? There is obviously a prominent role for judgements/beliefs in the way that a human makes sense of the things that appear to them. These judgements/beliefs concerning the way that reality is believed to be often bear little resemblance to the way that reality actually is. These flawed judgements both arise from, and feed the phenomenon of, human prejudice. So, whilst observation over time can lead to partial knowledge of the nature of things, the inability to observe the interior of things means that it can also lead to flawed beliefs, beliefs that a thing is a certain way (such as containing painful qualitative feelings) when it is actually another way.

These two implications mean that if things are to be known as they are, then the extremely limited ability of the visual sense to achieve this goal needs to be acknowledged. More than this, the visual sense needs to be seen as a barrier to attaining knowledge of things as they are due to fostering the phenomenon of human prejudice. Visual sensory systems are presenters of sense-originating things; they are moulders of ultimate reality. They also cannot penetrate through to the interior of things, they cannot provide knowledge of whether or not another 'human', or a 'cauliflower', or a 'chair', contains particular qualitative feelings. The realisation that this is so, the realisation that the visual sense can only provide a partial glimpse into the nature of things, through observation over time, is an accomplishment of rationality. Rationality has come to appreciate its inability to know things as they are through observing things.

The limits of rationality

The next question that naturally raises itself is: Can rationality provide knowledge of things as they are? Rationality itself can come to know its limitations, but it cannot provide knowledge of things as they are because it is an isolated bubble. Having said this, it is, of course, possible for a rational view to align with reality without having knowledge of reality. If one human rationalises that a thing has Quality X, and another human rationalises that this thing doesn't have Quality X, then one of them will inevitably have a view that aligns with reality, without this view constituting knowledge of reality. To make an educated guess, that happens to align with reality, isn't to have knowledge of reality.

In the face of the limits of rationality, the phenomenon of human prejudice, and the limits of the human visual sense, we are left with the following question: Is it possible for a human to attain knowledge of things as they are? Answering this question requires a consideration of mystical knowledge.

Mystical knowledge & the 'minimisation of rationality'

Humans acquire knowledge of the things around them through acquaintance with these things. You might recall that there are two types of acquaintance – observation over time and interaction. We have, so far, been concentrating on the first type of acquaintance, observation over time, and we have concluded that due to the fundamental limits of the visual sense, limits which are bolstered by the phenomenon of human prejudice, that humans only ever gain partial glimpses into the nature of things when they observe these things over time. So, knowledge of things as they are cannot be attained through observation; things as they are can be thought of as icebergs, and observation can only ever reveal their tips.

It is time to focus on the second type of acquaintance – interaction. To interact with a thing is simply to be in its vicinity; the closer the proximity between things, the greater is their level of interaction. An interaction between two things is a direct connection that is unmediated. In other words, things directly sense each other, but this sensing is not mediated by an elaborate sensory system, such as the human visual sense. As interaction is unmediated, free of human prejudice, and outside the realm of rationality, it provides a possible mechanism for getting to know things as they are. Indeed, this mechanism can give rise to a mystic.

A mystic is a human who has knowledge of things as they are, this knowledge arising from interactions with these things. Mystical knowledge could very easily be supposed to be the polar opposite of rational knowledge, the kind of knowledge sort by the scientist/philosopher. However, such an opposition would be a tad simplistic, given that the mystic is a creature of rationality. Mystical knowledge should not be straightforwardly opposed to rational knowledge. Having said this, it is likely, if not certain, that the 'minimisation of rationality' is a prerequisite for the existence of mystical knowledge. What does the phrase 'the minimisation of rationality' mean? If a human is dominated by thoughts, by ideas, by rational critique, by language, by formulae, by analysis, by facts and figures, dates and names, and labels, then they will be dominated by rationality. Being dominated by rationality is the opposite of the 'minimisation of rationality'.

Let us consider an example. Imagine that there are two humans who are located in a wonderful rainforest. One of these humans is an expert ecologist, they can name every tree, every plant, every insect, every animal; this human is full of knowledge, full of labels, full of rationality; their very being exudes knowledge and labels. The other human cannot give the name of a single tree, a single plant, a single insect or a single animal; however, they can deeply connect with that which is around them, they can connect with the forest and all its component parts; their essence exudes knowledge of things as they are, rather than knowledge of labels created by humans. The first of these humans is dominated by rationality, whereas the second exemplifies the 'minimisation of rationality'. Being dominated by rationality acts as a block to gaining knowledge of things as they are; being free of this domination, our second human will have much greater insight into the nature of things than our label-dominated ecologist. Our second human might even, let us suppose, have knowledge of things as they are; this would make them a mystic, a repository of mystical knowledge.

Lack of insight into the nature of things is a side-effect of education. Young children get force-fed labels, facts, figures, formulae, many then get degrees, and some even move on to get PhDs. As the scale of education progresses the human becomes a repository for an increasing amount of labels, of jargon, of complex terminology and human-created concepts. As the scale of education progresses the human becomes more and more removed from things as they are and becomes increasingly drawn into the superficial realm of human-created reality.

This does not mean that every human that is wholly devoid of education has knowledge of things as they are! The 'minimisation of rationality' is simply a prerequisite, a necessary condition, a seed, which can give rise to a mystic. And, of course, once a mystic has been born, they can then become educated, they can learn labels, they can get PhDs. However, it is surely hard to get overly excited by human-created labels and constructs if one has knowledge of things as they are! Why would one want to fill one's head with superficial concepts when one has access to the wonders of the cosmos! If you seek knowledge of things as they are, then be wary of the expert who throws around unnecessary jargon and complex terminology all over the place; such humans almost certainly do not have knowledge of the way that things are.

Let us delve a little further into the relationship between mystical knowledge and rational beliefs. It is obvious that that which is rationalised can be false. For example, it is well known that in the not too distant past there was a widespread human rationalised belief that the Sun revolves around the Earth, but we now know that this belief is false. In the past, humans have rationalised a plethora of things that any reasonable human alive today would consider to be unquestionably false. Whilst rational beliefs can be false, as human culture unfolds there is a tendency for false rational beliefs to be increasingly supplanted by less false rational beliefs and by true rational beliefs; this process leads to the advancement of rational knowledge. In contrast, mystical knowledge, by its very nature, is knowledge of things as they are.

We need to appreciate that mystical knowledge needs to be given expression through the rational faculty, through human words/concepts; after all, the mystic is a rational being. This expression will clearly be dependent on the present day level of development of the rational knowledge of the human species. What exactly does this mean? It means that a mystic that was born a couple of thousand years ago will express mystical knowledge in a form that is dependent on the extent to which human society, and human rational knowledge, has evolved up to this point. Whereas, a mystic that is alive today, in the year 2021, will express mystical knowledge in a more nuanced form.

Let us explore this a little further. The development of human rationality is intimately connected to the unfolding of our Solar System; more advanced rational knowledge entails a more highly unfolded Solar System. This means that the current day mystic has a dual advantage over their historical counterpart. The current day mystic is able to give expression to mystical knowledge in the light of increased human rational/scientific understanding of the nature of the Universe, and in the light of the fact that due to greater Solar-Systic unfoldment, much of what was mere speculation and hypothesis is now a past actuality which cannot be doubted. The contemporary mystic is thereby able to express mystical knowledge in a way that gives a much more detailed and comprehensive account of the Universe. This is because more has been revealed given that the evolution of our Solar System has progressed further, more of the nature of our unfolding Solar System has been revealed, become manifest, with concrete reality increasingly replacing future possibility. It is also because more has been revealed given that humans have observed and analysed the Universe much more, thereby providing an enhanced rational grounding for the expression of mystical knowledge. So, despite the fact that mystical knowledge is, by definition, true, it is also the case that a mystic is not isolated from the epoch into which they are born.

'Knowledge of a thing as it is'

We have been considering the extent to which humans can attain knowledge of the things that appear to them in their surroundings. If a human has fundamental knowledge of the nature of these things then they have knowledge of things as they are. We can now consider in a little more depth what it means to have 'knowledge of a thing as it is'. There are several aspects to this knowledge.

Firstly, 'knowledge of a thing as it is' is simply knowledge of a how a particular part of the Universe is at a particular moment in time. For example, a part of the Universe that you might call your 'pet cat', a part of the Universe that is currently sitting on your lap. If you have knowledge of the states that exist in this part of the Universe at a particular moment in time, then you will have 'knowledge of this thing as it is'. Due to the limits of the visual sensory system, in-depth knowledge of this thing, attaining 'knowledge of the thing as it is', requires mystical knowledge arising from interaction, a direct connection with the thing in the present moment. However, as we have explored, very limited knowledge of the thing can be gained through observation over time; in this way, one can gain some knowledge of the thing that is one's 'pet cat' via one's visual sense. A higher level of observations in the past will enable slightly more knowledge of the state of the thing to be attained in the present.

Secondly, 'knowledge of a thing as it is' is knowledge that the thing is not a static entity; 'knowledge of a thing as it is' is knowledge of a movement through time. In other words, to put it simply, staticness, non-movement, is an appearance which results from the way that our sensory organs interact with our surroundings. 'Knowledge of a thing as it is' is knowledge that that thing is a constant state of movement, of vibration, of flux, of becoming, of transformation.

Thirdly, 'knowledge of a thing as it is' is knowledge that that thing is not an isolated thing; it is ultimately not a real thing in its own right. The only 'thing' that is ultimately isolated, isolated in its own right, on its own terms, is the awareness of things. However, as awareness isn't actually a thing, but is rather a beholder of things, this means that all things interfuse into each other. In other words, what one might take to be isolated things are ultimately just movements without borders. The Universe is a vast labyrinth of intermingling borderless movements. Any borders, any divisions of these movements into individual things, any notion of things being static, are creations that are imposed onto the Universe by an observing entity.

' The way that things are' & the unobserved Universe

Having considered what it means to have 'knowledge of a thing as it is', we can now explore what it means to give an account of 'the way that things are'. Such an account involves intimate knowledge of the nature of the Universe and its constituent elements, which are the vibrating 'particles' that are the building blocks of the things that one observes around one. Such an account involves knowledge that the Universe is an intermingling unfolding entity which does not contain individual isolated things. Such an account involves an understanding of why the Universe has unfolded the way that that it has; in other words, it involves an understanding of the forces driving evolution at the level of human culture, at the level of the Earth, at the level of our Solar System, and at the level of the Universe. In short, such an account involves intimate knowledge of the unobserved Universe – the Universe in its unadulterated state, not one of the immense plethora of worlds that are created by diverse visual sensory systems. It goes without saying, that if one has such knowledge, then one has a pretty good idea how things are going to unfold in the future. If one has such knowledge, then one is a mystic.

Having knowledge of 'the way that things are' does not mean that one can predict every aspect of the future with certainty. For example, if one has knowledge of 'the way that things are', this doesn't mean that one knows who will win the Wimbledon tennis tournament in the year 2025, or who will be the President of the United States of America on the 17 March 2030. Knowledge of 'the way that things are' is knowledge of the fundamental forces and trajectories of our planet and our Solar System; if one has such knowledge then one inevitably understands the past, the present and the future. Knowledge of 'the way that things are' is not knowledge of the fine-grained superficial details of what will happen in the future; these details are irrelevant/unimportant and are not necessarily determinable in advance of their occurrence.

An account of 'the way that things are' is a comprehensive account of the nature of the unfolding Universe; it is an account which will not significantly change because it is fundamentally true. However, it is possible for particular aspects of such an account to get elucidated in greater detail. This is because whilst a mystic can see the big picture, through knowing the nature of the Universe, its journey through time, the meaning of life, the role of the human species in our Solar System, and other 'big' issues, there is inevitably scope for a deeper exploration of fine-grained issues within this account. It is also important to realise that a mystical account which arises at a later stage of Solar-Systic unfoldment could be thought of as a fresh interpretation of a previous mystical account. The greater unfoldment can enable more 'flesh to be put on the bones' of the previous account, with things able to be seen from a different perspective or in a new light.

The mystic's account of 'the way that things are', their mystical knowledge, might clash with contemporary rational beliefs, but this clash will not be irrevocable; the advancement of rationality can ultimately bring rational beliefs into line with mystical knowledge. Another way of putting this is to say that rational beliefs can lag behind mystical knowledge, resulting in a seeming conflict or contradiction, a conflict/contradiction that can be resolved through an advancement in rationality, a deeper penetration of the intellect into the nature of the Universe. The nature of reality, 'the way that things are', is not beyond the comprehension ability of rationality.

Doubting the mystic

It is possible that you might be sceptical about the existence of mystical knowledge. You might doubt that there is such a thing as a mystic. You might believe that it is impossible for a human to come to know 'the way that things are'. You might be convinced that the assertions of anyone who claims to be a mystic are nothing more than a personal opinion. All of this is perfectly understandable; after all, anyone could claim to be a mystic and to know the ultimate truths of the Universe. We face a simple brute fact here. There is no possible way for a non-mystic to know whether or not a mystic actually exists. Whereas, the mystic themselves, assuming that they exist, might not even know beyond any doubt that they are a mystic, a repository of mystical knowledge. The mystic might wonder whether or not their visions/experiences/feelings/insights are genuine. It is likely that some people believe that they are mystics who are in possession of mystical knowledge, whereas in reality they are not.

What is one to make of the possible existence, or non-existence, of the mystic? It is ultimately simply a matter of belief, one will either believe that mystics can exist, or one will believe that they are an impossibility. And, at the end of the day, it matters little whether or not you believe in the existence of mystical knowledge. In the big scheme of things, your beliefs are just not that important! After all, your belief, or lack of belief, in the existence of mystical knowledge is highly unlikely to make a meaningful concrete difference to the evolutionary unfolding of our Solar System. So, whether or not you believe in the existence of mystical knowledge is not particularly important.

Having said this, you might be interested in the big questions concerning human existence, the meaning of life and the nature of the Universe. If this is the case, then you might want to read the writings of those who claim to be a mystic, or the 'mystical' writings of those who do not claim to be a mystic, and see if they strike a chord with you, see if they inspire you, see if they make sense to you on some level. If they do, then you might come to believe the account that is presented of 'the way that things are'. If you are inspired, if you are affected in a positive way, if you live your life with more purpose and meaning and fulfil your potential to a greater degree, then that is what is important. At the end of the day, the mystic simply seeks to inspire you to see the wonders of existence, to maximise your potential and to live your life to the full; for, you were brought forth by the Universe out of the depths of nothingness precisely to do this. It matters little to the mystic whether or not you believe that they are a mystic. The mystic is a servant, not a deceiver, not a manipulator, not a seeker of fame, not an egotistical thing.

There are very good reasons to believe in the existence of mystical knowledge. After all, the information concerning how things are exists where those things are. So, why shouldn't this information be accessible to a human? A human is a thing, a part of the Universe, a repository of information; it is the way that it is, containing a plethora of states of interiority. Similarly, whatever a human observes is also a thing, a part of the Universe, a repository of information; it is the way that it is, containing a plethora of states of interiority. Why shouldn't these two things be able to connect in such a way that the information is shared? After all, the boundaries between all things are illusory and superficial, with the Universe being a vast labyrinth of intermingling borderless movements. This information sharing is simply a direct connection between two parts of the Universe; we have been referring to these direct connections as a type of acquaintance between things which we have called 'interaction'. As the Universe is deeply interconnected, information is widely shared between things; information sharing is the norm, rather than the exception. Furthermore, information concerning everything that has ever happened in the history of the Universe still exists; it is stored in the very fabric of the Universe. This means that if one can fully connect to the information within things that one can access information concerning the entire history of the Universe.

A question that raises itself is: Wouldn't these information-sharing connections between things make every human a mystic? Clearly, every human isn't a mystic! Why is this? We come back to rationality, having earlier explored the 'minimisation of rationality' in relation to mystical knowledge. Let us assume that all humans have connections to things, connections which contain information concerning the nature of those things. This means that every human is potentially a mystic. However, in the vast majority of humans their rationality acts as a barrier which causes them to not have access to this information. The mystic is the rare human who has 'overcome' their rationality in a way that enables them to plumb the depths of the information that exists both in themselves and in the things that surround them. This 'overcoming' is not a result of conscious striving, it is the outcome of a life lived in a particular way; it is more like an accidental side-effect than a sought after goal. In the mystic, 'the way that things are' reveals itself, becomes apparent, and this revelation becomes interfused/synthesised with rationality in a way that results in the expression of mystical knowledge.

The universal pursuit of ecstasy

If we seek a pithy phrase to describe 'the way that things are', then our phrase is: 'the universal pursuit of ecstasy'. What on earth does this mean?! The word 'universal' refers to every part of the Universe. The word 'pursuit' refers to the 'inner states' of these parts, and the way that these states change from moment to moment, continuously qualitatively transforming themselves into new states; every part of the Universe is a qualitative feeling that is in pursuit of more preferable feeling. The continuous transformation of these states, this process of change, gives rise to what we might want to call 'evolution'; that is, evolution of our Solar System, evolution of the Earth, evolution of life, evolution of individual life-forms, evolution of human culture; that is, evolution of everything. The word 'ecstasy' refers to the state which these parts are seeking to attain, the state which they are striving after; it is the end-point of everything; it is the goal which is universally sought; it is the state of maximal excitement and exhilaration.

This is an appropriate time to say something of great importance concerning the distinction between the 'inner' and the 'outer'. In describing the universal pursuit of ecstasy I used the phrase 'inner states'. This is because we, as humans, are so accustomed to thinking about things as having an 'inner state / existence' and an 'outer state / appearance'; however, in reality, there is no division in things between 'inner' and 'outer', there are just states. The human idea of an inner/outer separation is fostered through human visual perception, which is the key way in which humans make sense of their surroundings. Human visual perception can only perceive the 'outer' of things, but the human intellect can easily rationalise that there is so much more to some of the things that they visually perceive, than they are able to visually perceive, so an 'interior' is postulated in addition to the external appearance. This 'interior' is taken to be the home of things such as thoughts and feelings and the awareness of things. This postulated 'interior' is actually 'the way that things are'; it is the real fundamental reality; the Universe is an intermingling web of states of feeling, thought and awareness. So, the phrase 'inner states' refers to this intermingling web, which actually has no division between 'inner' and 'outer'. What humans typically take to be the 'external' part of things is a superficial creation of the human visual sensory system.

When considering the universal pursuit of ecstasy, the processes/states that exist in all parts of the Universe, it is useful to keep in mind the natural tendency towards human prejudice. This is because human prejudice can easily act as a mental straitjacket which blinds one to the existence of the universal pursuit of ecstasy. Also, it is useful to keep in mind that different parts of the Universe have differential capacities to attain ecstasy. What unites all parts of the Universe is their pursuit of ecstasy, which is a pursuit that results in the coming into being of states that more closely approximate the state of ecstasy, whenever and wherever circumstances allow. What these states actually feel like, in terms of both their depth/intensity of feeling and the qualitative nature of this feeling, will vary from part to part, from place to place, from thing to thing, because each thing is at its own stage of the journey towards ecstasy. Whilst the entire Universe is on the same journey, the overwhelming majority of the Universe will never get anywhere near its desired ultimate destination.

Let us explore this a little further. The state of ecstasy is a state which everything, every part of the Universe, seeks to attain. However, the state of ecstasy is not a state which any part of the Universe can suddenly jump to; it is not a state which is easily attainable. The vast majority of the Universe is on a long journey towards the state of ecstasy; ecstasy is the ultimate goal which requires an immense multitude of hurdles to be overcome before it might possibly be attained. The term 'long journey' should here be thought of in cosmic terms, not in human terms. For a human, a short journey might be a 30 minute commute, whilst a long journey might be a 16 hour airplane flight. In cosmic terms, a long journey needs to be thought of in terms of millions and billions of years! Another way of thinking about this is to equate the evolution of the Universe with the universal pursuit of ecstasy; they are one and the same thing.

An example to illustrate all of this might be useful. Cast your mind back to the time when the wheel was invented; from this moment in time it was, from the perspective of human history, a relatively long journey to get to the bringing forth of the formula one sports car. It was obviously not possible to jump straight from one state to the other state; it was not possible to jump straight from the state of 'wheel invention' to 'formula one sports car invention'. Rather, the state of 'wheel invention' was part of a journey which would ultimately lead to the bringing forth of the formula one sports car. Without the invention of the wheel, the bringing forth of the formula one sports car would clearly have been impossible. This end result required a long process of transformation and the bringing forth of a plethora of intermediate states. This example is useful because it reinforces the point that the attainment of something can be a long drawn out journey, rather than something which can suddenly be attained by a click of the fingers.

The various parts of the Universe are on a journey towards ecstasy, just like there was a journey from 'wheel invention' to 'formula one sports car invention'. The goal of ecstasy is a common goal, but the various parts of the Universe are at very different stages of the journey towards ecstasy. Don't confuse 'being on a journey' with 'getting to a destination'. The destination provides the direction of travel, but if conditions don't allow the journey to continue, then a thing can be stuck in a perennial traffic jam. What this means is that the actual states that are currently being attained by various parts of the Universe, as they strive for ecstasy, moment by moment, are very different. The states that are attained are all on the path to the attainment of ecstasy, they all increasingly approximate the state of ecstasy, yet they vary greatly in the degree to which they approximate ecstasy.

This description of the journey towards ecstasy which various parts of the Universe are progressing through is a useful way of putting 'flesh on the bones' of the notion that different parts of the Universe have differential capacities to attain ecstasy. A part of the Universe that is towards the beginning of the journey towards ecstasy has a very low level of feeling, and thus a low capacity to attain ecstasy. In striving for greater feeling, a greater level of excitement/exhilaration, these parts of the Universe, if their striving is successful, bring forth states which are very slightly closer to ecstasy, whilst being a million miles from ecstasy itself. In turn, these new states which are brought forth themselves strive for greater feeling, a greater level of excitement/exhilaration, and if they are successful they will transform themselves into states which are ever so slightly closer to ecstasy, yet which are still a million miles from ecstasy itself.

The journey of the Universe to ecstasy can be compared to the construction of a building. Let us imagine a block of flats. On the ground floor there are small and fairly standard flats, as the floors of the building are ascended the flats become more spacious and extravagant, and on the top floor of the building is the most magnificent and majestic penthouse apartment that is imaginable. The penthouse apartment is obviously the state of ecstasy. After this penthouse apartment has been constructed, it can be made even more majestic through extravagant furniture and dazzling works of art. Analogously, once the state of ecstasy has been attained, it can become even more ecstatic.

If we go back to the time before the building was constructed, when there was just a bare patch of ground, then we can think of this as the beginning of the Universe. When the foundations of the building were being laid this can be thought of as parts of the Universe which are a million miles from ecstasy striving for ecstasy. As the construction of the building proceeds, as walls are erected, windows are inserted, plumbing is installed, walls are painted, carpets are laid, and furniture arrives, this can be thought of as the universal journey towards ecstasy. What can we learn from this comparison? When the finished penthouse apartment, the state of ecstasy, has come into being, it exists alongside a plethora of non-ecstatic states which have helped to bring it into existence. The penthouse apartment exists, but it would not exist if it were not for the flats on the lower floors; it would not exist if the building did not have foundations. If the flats on the second floor were suddenly to be removed then the penthouse would come crashing to the ground and would cease to exist. The penthouse is the state of ecstasy, but it depends for its existence on states which have previously existed, and which currently exist, which are non-ecstatic. When the Universe reaches the state of ecstasy, not all parts of the Universe reach this state; indeed, the vast majority of the Universe is inevitably the 'foundations' which enable a few parts of the Universe to attain ecstasy. Of course, these 'foundations', whilst supporting the existence of ecstasy, are themselves striving to become ecstasy.

So, all parts of the Universe are striving for ecstasy, but only a few parts of the Universe will be lucky and actually make it. One could think of this in terms of the foundations of the building and the flats on the lower floors sacrificing themselves so that the state of ecstasy can come into existence in the form of the penthouse. However, there is no sacrifice. There are just states of the Universe which are striving for ecstasy as best as they can and some of these states will inevitably be further along the journey than others. Think of humans running a marathon. These humans are all striving to complete the course and to run it as fast as they possibly can. However, it is possible that not all of the runners will finish the race, and it is inevitable that some of the runners will complete the course in a much quicker time than others. The vast majority of runners have no realistic chance of winning, yet they are happy to strive to the best of their ability, and they are happy to see another runner be victorious. The vast majority of the runners are the support act to the winner, yet without a field of competitors there can be no winner. So, the entire field of runners is of great value, whilst the winner is the most precious treasure. If we switch our focus from marathons to sprinting, then, in recent years, the most precious treasure is known as Usain Bolt; he is magnificent and majestic, like our penthouse. Yet, if he had no competitors to run against, he would be nothing. Analogously, the state of ecstasy is only a precious treasure because of its relationship to non-ecstatic states of the Universe.

The pursuit of ecstasy is universal; the pursuit exists in all parts of the Universe. What this means is that when one comes face-to-face with the universal pursuit of ecstasy that one will gain insight into the way that the Universe changes through time; in other words, one will appreciate the forces that underpin the evolution of all parts of the Universe. The universal pursuit of ecstasy results in processes of change through time which we can look at and call 'evolution'. One can look at these processes of change/evolution from 'the outside' and seek to understand them through taking objective measurements and through conducting studies and experiments. One can seek to attach objective labels to various parts of this universal journey, this process of striving and unfolding. All of this can sound quite respectable, couched as it is in academic circles, with impressive sounding labels and names. However, such objective analysis can never result in an understanding of why things evolve the way that they do. Such an understanding requires one to appreciate the 'internal' states of things, which are states which cannot be scientifically analysed and academically labelled. If one penetrates to the 'interior' of things, then one will know all that one needs to know. One will know that things evolve the way that they do because all things are driven by the universal pursuit of ecstasy.

It is important to realise that the universal pursuit of ecstasy does not cease with the attainment of ecstasy. When the state of ecstasy has come into existence, the Solar-Systic journey that is the universal pursuit of ecstasy continues; it continues both within that which is non-ecstatic and within the realm of ecstasy itself. Whilst ecstasy is the state of maximal feeling that a thing can attain, states of ecstasy themselves vary in their intensity; following the coming into existence of ecstasy, there is still a striving for more preferable feeling.

Question: Why do particular atoms interact so as to form particular molecules?

Answer: The universal pursuit of ecstasy.

Question: Why do planets form?

Answer: The universal pursuit of ecstasy.

Question: Why did life arise on the Earth?

Answer: The universal pursuit of ecstasy.

Question: Why did the human species evolve?

Answer: The universal pursuit of ecstasy.

Question: Why did the human species explore the Universe in great depth and bring forth advanced technology?

Answer: The universal pursuit of ecstasy.

Question: Why do life-forms procreate?

Answer: The universal pursuit of ecstasy.

Question: Why do comedians exist?

Answer: The universal pursuit of ecstasy.

Question: Why do humans kill each other and carry out horrific acts of cruelty?

Answer: The universal pursuit of ecstasy.

Question: Why do dolphins jump out of the ocean and why do they help humans in danger?

Answer: The universal pursuit of ecstasy.

Question: Any question of any real significance or importance.

Answer: The universal pursuit of ecstasy.

It is probably helpful not to jump to any conclusions or to analyse the above questions and answers too much at this stage. After all, a little knowledge is a fertile breeding ground for grave misunderstanding. There is much more to be said concerning the universal pursuit of ecstasy, and the meaning and significance of these questions and answers will only become clear as our journey progresses. For now, one can simply appreciate that the universal pursuit of ecstasy is continuously in operation throughout the entire Universe. This means that it is in operation within oneself and within everything that one observes in one's surroundings. There is no place that is not the home of the universal pursuit of ecstasy.

Increasingly intense pleasantness

The universal pursuit of ecstasy is a pursuit that involves the striving for feelings that are pleasant, as opposed to unpleasant, and that are ideally also more intense. A higher intensity of pleasant feeling is the most desirable form of pleasantness. We can refer to this process, the universal pursuit of ecstasy, as the striving for 'increasingly intense pleasantness'. This phrase encapsulates the desire of all things in the Universe to experience both feelings that they find to be pleasant, as opposed to unpleasant, and to feel more deeply, more intensely. When a thing attains 'increasingly intense pleasantness' it will have a heightened level of excitement/exhilaration.

If the opportunity arises, as a thing interacts with its surroundings, for that thing to experience 'increasingly intense pleasantness', then that feeling is grabbed and held onto, whilst any feelings that are experienced that are unpleasant are pushed away. This 'grabbing and holding onto' involves a thing expanding in size through transforming itself into a new thing; whilst, 'pushing away' involves a thing maintaining its size. What exactly it means to talk of a 'thing' in this context will only become clear as our journey progresses. For now, we can imagine that the entire Universe is represented by a sealed room and that within this room there are a plethora of tiny balls. These balls are continuously bouncing around the room, bouncing off the walls of the room and bouncing into each other. In terms of the universal pursuit of ecstasy, each ball represents a part of the Universe, which means that it is a particular feeling state. If, as two balls approach each other, they find that they are experiencing 'increasingly intense pleasantness', then they grab and hold onto each other; in this way they become a new thing. If, as two balls approach each other, they find that they are experiencing unpleasantness, then they push each other away. If, as two balls approach each other, one of them experiences 'increasingly intense pleasantness', whilst the other experiences unpleasantness, then there is a struggle and the ball which has the most intense feeling, the deepest level of feeling, will prevail. These diverse experiences are the type of acquaintance between two things that we are calling 'interaction'. As two balls get closer to each other their level of interaction with each other increases.

Whilst our balls are bouncing around they will experience a diverse range of feelings that they find to be both pleasant and unpleasant, but there is an inevitable movement through time in the direction of 'increasingly intense pleasantness'. However, we need to appreciate that once things attain a higher intensity of pleasant feelings there is a price to pay, for, they then leave themselves vulnerable to experiencing unpleasant feelings of a higher intensity. The intensely unpleasant is the price that needs to be paid for the attainment of the intensely pleasant.

Human purpose

Having outlined the universal pursuit of ecstasy, it is time to say a few words concerning human purpose. After all, this section is called *Human Purpose & the Universal Pursuit of Ecstasy!* Given that all parts of the Universe that are not ecstasy are striving for ecstasy, are on a journey towards ecstasy, there is an obvious sense in which all parts of the Universe have a purpose. The purpose of any part of the Universe that is not ecstasy is to move towards ecstasy, through attaining 'increasingly intense pleasantness'. This is an acceptable use of the notion of 'purpose'; there is no need for awareness, deliberate design, or anything else, for there to be purpose.

As a part of the Universe moves in the direction of ecstasy, towards ecstasy, it becomes increasingly excited/exhilarated. This movement itself is the fulfilment of a purpose for two reasons. Firstly, it is a step closer to the ultimate goal of the attainment of ecstasy. Secondly, in itself it brings the Universe into a more preferable state; for, the more states of heightened excitement/exhilaration that exist, the better the state the Universe is in. In other words, more intense feeling/vibrancy in a particular location, and in as many locations as possible, is preferable to deadness and dullness.

As we explored, when comparing the journey of the Universe towards ecstasy with the construction of a penthouse apartment, the goal of attaining ecstasy will not be achieved by all parts of the Universe. The purpose of all non-ecstatic parts of the Universe is to strive towards ecstasy, which for the vast majority of these parts entails not actually attaining ecstasy. Given this, you could reasonably assume that the purpose of the human species is to help the Universe move towards ecstasy, but this assumption would be wrong. For, the human species is the state of ecstasy! The human species is that part of the Universe which has been striven for over millions and billions of years. Why is this? What is so special about the human species? To be clear, we are talking about concrete reality, about states of ecstatic existence, states of qualitative feeling which exist in the things that we call humans. You might be wondering why the states of the Universe which are located in humans are ecstatic, whilst the states of the Universe which are located in things like chimpanzees and dolphins are not ecstatic. All will become clear as our journey progresses.

The first question before us is: What is the human species? In the context of a very slowly unfolding and evolving Universe, which is gradually transforming itself in the pursuit of ecstasy, and in which there are no independent isolated 'things' (the independent objects which humans typically take to exist being creations of the human visual sensory system), what does it mean to talk of the 'human species'? And: When exactly did the 'human species' come into existence? When one has the answers to these questions then one will be well on the way to understanding human purpose. One will also be well on the way to understanding why the human species is the state of ecstasy. For, human purpose and the state of ecstasy are so inextricably linked, that they are, to all intents and purposes, one and the same thing. Whilst the human species is ecstasy, some states of ecstasy are more ecstatic, more intense, more developed, than others. The purpose of the human species is to both move the Universe to more pervasive and more intense ecstatic states and to increase the pervasiveness of highly excited/exhilarated states in the Universe; in this way, the state of feeling in the Universe is maximised.

What is the human species?

It is easy to believe that this question has a simple and obvious answer. When one considers the world of things, the world of objects that appears to one, then dividing these things into those which are 'human' and those which are 'non-human' seems to be simplicity itself. Biologists also provide a simple definition of a species: life-forms which are able to breed and produce fertile offspring are members of the same species. Things aren't as simple as this though! This biological definition of a species has shaky foundations because it turns out that two things that we would want to consider to be from very different species can breed and produce fertile offspring! There are actually a plethora of different ways in which one can define what a 'species' is; this, in itself, raises alarm bells and raises the prospect that in reality there is no such thing as a 'species'. More fundamentally, we have come to appreciate that the things that we observe around us are things that have been moulded and segregated by our visual sensory system, which means that to talk of a 'species' of things is to talk about a human-created concept.

So, what we usually think of as a 'species' of things, is something that exists in the human conceptual realm. One day we can identify a group of things and assert that all of these things are part of 'Species XYZ'. The next day we can assert that, actually, this same group of things contains two different species – 'Species XYZ' and 'Species ABC'. We are not interested in such fleeting categorisations; we are interested in the fundamental nature of reality. We are interested in the question: Does it make any sense to say that there are 'species' in the unobserved Universe ('the way that things are')?

Let us start with a related question: When our perspective is the fundamental nature of reality, 'the way that things are', what does it mean to talk of the human species? Let us approach this question with a few answers; answers which are all true. The human species is the state of ecstasy. The human species is the zenith of the evolutionary progression of life on Earth. The human species is the zenith of the evolutionary progression of our Solar System. The human species is the saviour of life on Earth. The human species is the penthouse apartment, other life-forms in our Solar System are the apartments on the lower floors, and the non-living parts of the Universe are the foundations of the building. The human species is the state of maximal universal excitement which all parts of the Universe seek to be, but which very few will become. The human species is that part of the Earth which is scientific and technological. The human species is that part of the Earth which is the master modifier of its surroundings. The human species is that part of the Earth which is thrust out in opposition to the rest of the Earth, so that it can see its opposite as the other, as that which is to be used and abused, and through such using/abusing the human species both suffers immensely itself and brings suffering to that which is not itself. The human species is the 'slave', the 'servant', of life and our Solar System. The human species is a collection of 'cosmic puppets'.

So many answers to the same question! Some of these answers will surely be blindingly obvious to you, even though you might not have previously grasped their cosmic significance. Whilst, there is a good chance that you will also find some of the answers to be rather opaque. We will be exploring these answers throughout our journey; for now, we can start to consider the blindingly obvious answers: *The human species is that part of the Earth which is scientific and technological. The human species is that part of the Earth which is the master modifier of its surroundings.* You don't need to be a mystic to be able to easily realise that this is so! However, let us put a little more 'flesh on the bones'. The entire progression of the Universe from its beginnings to the current day is a journey; it is a journey that is aptly referred to as the universal pursuit of ecstasy. This journey, this pursuit, can also be thought of as a journey of modification of the other (of that which is not itself). Modification of the other is the ability of a part of the Universe to interact with another part of the Universe in order to transform itself, thereby bringing into being a higher state of excitement/exhilaration. It is about forging connections with other things; things coming together and forming new things. So, atoms modify their surroundings to form molecules, molecules modify their surroundings to form cells, cells modify their surroundings to form organisms, humans modify their surroundings to create mobile phones, and so on. Every thing modifies many other things. This modification is an automatic process which involves things doing what feels right; things naturally want to move to 'increasingly intense pleasantness', a higher state of excitement/exhilaration, a higher state of aliveness, a higher state of stimulation, a higher state of vibrancy. This process results in the unfolding of the Earth, the unfolding of our Solar System, and the unfolding of the Universe.

As this unfolding process extends through time something quite amazing, but obvious, occurs: the ability of things to modify their surroundings increases. So, to put it rather simplistically, eukaryotic cells are things which have a greater modification ability than the non-living things that preceded them, plants are things which have a greater modification ability than eukaryotic cells, animals are things which have a greater modification ability than plants, and as time passes animals transform themselves into new forms which have a greater modification ability, and this process culminates in the coming into being of the human species, the master modifier. When one appreciates that the entire journey of the Universe is a journey of the increasing ability to modify, and one appreciates that the human species is the master modifier, then one can easily see that the human species is the zenith of the evolutionary progression of life on Earth. We can also say that the human species is the zenith of the evolutionary progression of our Solar System, or of the Universe. This is not to say that there might not be other master modifiers in other distant parts of the Universe. For, the human species is the master modifier; so, wherever master modifiers exist, the human species exists.

This is an important point to appreciate. The human species is not a particular appearance, not a collection of things which typically has a particularly bodily form; and, it is not a particular set of genes. The human species is a term which refers to a particular activity, the activity of master modification. Wherever master modification exists the human species exists. Such a conclusion is nothing more than a return to where we started, the realisation that things are not defined by their appearance to a particular observer. Rather, things are defined by their movements, their actions, by the way that they interact with other things. Recall that one gains no knowledge concerning a thing from a static appearance, but that observation over time, observation of movements/actions, can generate some knowledge of the nature of a thing.

So, whilst in reality there are no such things as 'species', there is a collection of activities which answers to the label 'master modification', and it will be fruitful for us to sometimes use the words 'human species' to refer to this collection of activities. If it makes any sense to use the term 'species' to refer to parts of 'the way that things are', then to talk of a 'species' is to talk of a collection, a 'family', of similar movements that exist in different locations. Can you visualise what these words are attempting to convey? Imagine the particular parts of the Universe that humans call 'butterflies'. Imagine the movements that a particular 'butterfly' makes as it moves, the movements that it makes as it goes through its life. Any 'thing' that makes movements like these deserves to be called a 'butterfly'. If you visualise all of these particular movements that currently exist in the Universe, then where these movements exist, 'butterflies' exist.

The unfolding journey of increasing modification ability towards the human species obviously means that many non-human things on the Earth will be able to use and design tools to varying degrees of sophistication. What makes the human species the master modifier is that the human species is that part of the Universe which has gone through a scientific and technological revolution. The ability to bring forth technology, rather than to bring forth tools, is the hallmark of the master modifier. The human species is the master modifier. So, it obviously follows that the human species came into existence when the scientific and technological revolutions occurred. If we are looking for an exact moment in time when the human species came into existence on the Earth, then it is the moment when the first piece of technology was brought forth into existence. Before this moment tools were used, and many of these tools were very advanced, but they were not technology, so the human species did not exist. For sure, 'things' existed that had the appearance of a human; but, these 'things' were just 'animals', 'non-humans', 'non-master-modifiers' using tools.

What distinguishes an advanced tool from technology? What was the specific day/year that the human species came into existence? Such questions could be debated long and hard. Yet, at the end of the day, they do not need to be answered. We just need to be able to clearly see that there is a long progression from tool use, to advanced tool use, to scientific revolution, a revolution which enables the bringing forth of technology. We could say that the difference between technology and an advanced tool is that technology has been designed with knowledge gleaned from science. Then the questions that raise themselves are: What exactly is science? Did science precede the 'scientific revolution'? Perhaps it makes sense to say that it did. We could go round and round in circles, but thankfully there is no need. All we need to appreciate is the process that has occurred on the Earth. We have no need to give an exact date for the birth of the human species. We have no need to give an exact definition of science. And we have no need to give an in-depth account of the difference between an advanced tool and technology; however, the explanation of this difference just given seems to be perfectly adequate.

What is life?

We have been exploring the question of what makes the human species the human species. In so doing we seem to have jumped ahead of ourselves a little! If one is to properly appreciate the purpose of the human species, its cosmic purpose and significance, then one needs to appreciate the difference between the living and the non-living parts of the Universe. So, it is time to explore the question: What is life?

You probably think that you have a pretty good understanding of which parts of the Earth are living and which are non-living. When you are walking through a park you encounter easily identifiable living and non-living things. The humans you see walking in the park are living, so are the dogs that are running in the park, as are the birds in the trees, the trees, the squirrels, the worms, the cats, the shrubs, the grass and the flowers. In contrast, the wooden fence is non-living, so is the lead which connects dog to owner, and the benches, the rubbish bins, the street lamps and the concrete paths. A division between the living and the non-living along these lines is surely correct. There is sometimes a lot to be said for common sense!

However, there are lengthy and ongoing debates concerning what exactly life is. One reason for this is that there are borderline cases where it is questionable whether a thing is living or not. There is also a deeper issue arising from the possibility that the division between the living and the non-living is a division that humans project onto the Universe, rather than a division in the Universe itself. Thinking along these lines results in some people denying that there is actually a division between the living and the non-living; these people typically believe that the entire Universe is living, because the alternative is to believe that they themselves are dead! Indeed, you might well have heard the view expressed that 'everything is alive'. You really shouldn't be tempted by this line of thought. It is surely obvious to you that there is a real division in the Universe between the living and the non-living, a division which exists independently of human observation, thought and classification.

Let us consider the living/non-living division from the perspective of our evolving Solar System. If we go back far enough in time our Solar System was in a wholly lifeless state; there was no division between the living and the non-living because our Solar System was wholly constituted by the non-living. Then, when conditions allowed, the striving of the non-living to move towards ecstasy resulted in it transforming itself into the living; our evolving Solar System had brought forth life. Life has arisen on non-Earth locations, locations such as Venus, where the conditions enabled life to arise, but where the conditions were not suitable for life to thrive. In such locations a temporary chasm opened and then closed. In other words, in these places the universal pursuit of ecstasy was thwarted; it was halted in its tracks. How sad! Venus moved from A) Lifeless to B) Living/Non-living and on/back to C) Lifeless. How terribly sad!

On the Earth the chasm opened, life was brought forth, and as the conditions were suitable for life to thrive the chasm has remained open; there is still, in the year 2021, a division on the Earth between the living and the non-living. On the Earth the universal pursuit of ecstasy has been able to move forward, to flourish, to flower, to reach its ecstatic human pinnacle. How wonderful! In the future, if we extend far enough in time, eventually the chasm will close; this will happen because of the expiration of the Sun. This is not to say that the universal pursuit of ecstasy will have been fatally impaled. For, by this time, life will have fled the Earth and moved to new homes in other solar systems. In other words, ecstasy will have populated other solar systems, and will have taken other life-forms with it to their new home. This might seem to be far-fetched, but remember that we are talking about a very long time in the future. Just keep in mind how far technology has advanced in the past 100 years, then talk of such events in hundreds of thousands of years will seem to be far from far-fetched.

This way of thinking about life, as an entity which is brought forth when a chasm opens in the Universe, but which goes out of existence if the chasm closes (if the universal pursuit of ecstasy is thwarted) is useful because it helps one to grasp the idea that life is fundamentally a single entity. Life is that part of the Earth which is not the non-living. Life is that part of the Earth which wouldn't exist if the Earth fell back to a state of lifelessness. The Earth can be divided sharply into two parts – the living and the non-living. We can describe the living part as 'life on Earth' or as 'planetary life'. These terms refer to the part of the Earth that is alive. At a broader level, the term 'life' can be thought of as referring to life wherever it exists in the Universe. Whilst, at a narrower level, 'life on Earth' can be divided into two parts – 'complex life' and 'non-complex life'. 'Non-complex life' is a term that refers to extremely simple life-forms: bacteria, single-celled life-forms and other microbes. These life-forms can exist in extreme conditions and can therefore exist when life on a planet is not thriving. The division is important because when life is firmly established on a planet it evolves 'complex life' which is a sign that life is thriving and that the universal pursuit of ecstasy is in full swing. If you find life-forms on a planet such as cats, dogs, bats, dinosaurs, trees, elephants, dolphins, roses, carrots and birds, then life on that planet is thriving. Fantastic!

Thinking of planetary life in this way, as a single entity, or as a single entity with two parts, is not how most people typically think about life. Most people typically think about life in terms of individual organisms. In other words, individual organisms are conceptually isolated, and then aggregated, so as to form life. According to this way of thinking, life on the planet at a particular moment in time could be thought of as being constituted by, say, 51678 dolphins + 2330701 ladybirds + 605030327 humans + 7068005638 trees, and so on. If you can give up this individualistic and aggregative way of thinking about life, and move to a visualistic and holistic way of thinking, then this will aid your understanding. Try and visualise life on Earth as a single interconnected entity which spans the planet; the birds in the sky, the worms in the earth, the fish in the sea, the humans in the cinema – all limbs of a single entity. This entity is 'life on Earth' / 'planetary life'. 'Planetary life' isn't simply a collection of individual life-forms; it is a single entity that has parts in multiple locations. If this distinction isn't clear to you it may help to think of your body; you probably think that your body isn't simply a collection of individual parts, rather, it is a single entity which could be conceptualised as being constituted out of individual parts.

What is of cosmic significance, of real importance, is the state of the single entity that is 'planetary life'. The interests of a particular 'thing' – a dog, a whale, a bird, a human – are not of real importance in the big scheme of things; these individual interests have no cosmic significance. The interests of what we think of as a 'species' are also of little importance, of little cosmic significance, unless we are talking about the human species (ecstasy) or a species (a 'family' of similar movements) which is of essential importance to the existence of ecstasy. The only thing that is important is the existence and thriving of the living whole – 'planetary life' – as it strives to bring forth and maintain the state of ecstasy.

Let us briefly return to the view that the division between the living and the non-living is a division that humans project onto the world, rather than an actual division in the Universe itself. It seems obvious that there is a real division in the Universe itself between the living and the non-living, and it seems obvious that the living is the more precious of these two parts. However, it also seems obvious that this distinction is a subtle one. The living part of the Universe is a heightened state of 'excitation' or 'exhilaration'. In other words, a commonality between all things that are living is that they are at a higher level of excitation/exhilaration than all things that are non-living. That is to say, all living things are further along the journey towards ecstasy than all non-living things.

The non-living

Having started to explore the difference between the living and the non-living, we really need to say more about the nature of the non-living. We have seen that the living is a portion of the Universe that is further along the journey towards ecstasy than the non-living. Living states are distinguished from non-living states through their higher level of excitation/exhilaration. Another way of putting this is to say that living states have a greater modification ability than non-living states. These are one and the same thing; for, a greater modification ability equates to greater excitation/exhilaration. As a thing attains a greater modification ability it simultaneously attains 'increasingly intense pleasantness'. This means that things instinctively strive for a greater modification ability. We can think of the living/non-living division in terms of metabolism. Living things are distinguished from non-living things through the fact that they metabolise; that is, they draw parts of the Universe that are not themselves into themselves, and then excrete things back out from themselves into the Universe. For example, humans inhale oxygen and exhale carbon dioxide; they also take food and water into themselves and then release faeces and urine back to the Universe. In contrast, the things in the non-living Universe simply bind to other things, form connections with other things, without taking other things into themselves. From the perspective of the universal pursuit of ecstasy, we can see that metabolism is a state of enhanced modification ability, a state of heightened excitation/exhilaration that is the hallmark of the living.

The non-living is itself a state of excitation/exhilaration, a state of low excitation/exhilaration which continuously strives to move towards a higher state of excitation/exhilaration. Such striving will be successful if the surrounding conditions are suitable. A hallmark of non-living things is their ability to only modify other things through binding to them, through forming external relationships; they are unable to modify things to a higher degree by drawing them into themselves through metabolism. Whereas all living things are closer to the state of ecstasy than all non-living things because of their greater modification ability in being able to draw other things into themselves through metabolism, rather than only being able to externally modify things. Living things modify other things both by drawing them into themselves and by initiating external modifications. The human species is that part of life that is ecstasy because it is the master modifier in terms of both drawing things into itself and modifying that which is external to it.

It is clear that the non-living can be thought about from two perspectives. One can think about the non-living in terms of its lack of ability to metabolise, its lack of ability to modify other things to a greater degree by pulling them inside itself. This is clearly an external perspective, and can be thought of as a 'scientific' perspective; it says very little about what is actually going on from the perspective of the non-living itself. The second perspective is the internal perspective, what it is actually like to be a non-living thing, and why the non-living thing does what it does. This second perspective is the realm of states of excitation/exhilaration, the realm of the universal pursuit of ecstasy. It is this second perspective which we will explore further now.

You might be sceptical that states of excitation/exhilaration exist in all non-living things. You might think: Why should I believe in something that there is no evidence for? The reality is that there never will be any evidence for such states, just as there will never be any evidence that another living thing, such as another human, contains states of excitation/exhilaration. Despite this lack of evidence, I assume that you believe that other humans contain states of excitation/exhilaration. The athlete that crosses the finishing line first to win an Olympic gold medal; the student who learns that they have passed their exams; the pride of the parents when their child gets their first job; the artist who is in a state of flow as they bring forth their masterpiece; any human who is vigorously engaged in an activity that they are passionate about. Do you believe that these humans contain states of excitation/exhilaration? Do you believe that all humans contain states of excitation/exhilaration? You have no evidence that such states exist, no conclusive proof, but you surely believe that they exist. Why might you believe that states of excitation/exhilaration exist in some things, but that they do not exist in all things? Such a belief is in accordance with the phenomenon of human prejudice which we explored earlier. Why does human prejudice exist? There is much more to be said about this, and we will be in a position to return to this important question later in our journey.

Why might you come to believe that states of excitation/exhilaration exist in all non-living and living things? There are two routes to such a belief. Firstly, you might have direct knowledge of the nature of things / mystical knowledge. If states of excitation/exhilaration exist in all things, and one has a direct mystical connection with non-living things, then one will know that such states exist in non-living things. Secondly, you can get to such a belief through a process of rationalisation. So, you might rationally believe that everything that exists has come into existence through a very gradual process of evolution which is traceable back to the event known as the 'Big Bang', a time when only non-living things existed. You might also rationally believe that states of excitation/exhilaration exist in all humans, all animals, and even all living things. Given these two beliefs, one might find it hard to make rational sense of the possibility that at a certain stage of the gradual evolutionary unfolding of the Universe that states of excitation/exhilaration suddenly sprung into existence out of that which is wholly devoid of such states. You might find it to be much more plausible to believe that as the Universe has gradually unfolded and evolved since the Big Bang that there has been a modification and evolution of the qualitative states of excitation/exhilaration that have always existed, rather than a sudden springing into existence of such states out of a sea of nothingness.

Through either of these two routes one can end up face-to-face with the phenomenon of human prejudice – one fully sees it, one sees through it, and one comes to appreciate things as they are. If you have attained this level of appreciation, then you will, no doubt, spend time observing other humans, observing how they act, observing what they say, and you will conclude that human prejudice is currently extremely widespread in the human species. There is also a chance that you will have come to appreciate the cosmic importance, the cosmic necessity, of the current pervasiveness of the phenomenon of human prejudice.

What can we say about the nature of the states of excitation/exhilaration that exist in all non-living things? The first thing that we can say is that these states feel a certain way, they have a qualitative element. From one's own perspective, one will be aware that one's body contains a plethora of such states at any moment in time. These states can vary greatly in nature, from buzzing with excitement, to mild tingling, aching and severe pain; these states are unified by the fact that they feel a certain way; they have a qualitative element. Sometimes these states are more vivid and pronounced than at other times, but they always exist. The qualitative states that exist in non-living things will be very different to the qualitative states that exist in a human. After all, non-living things and humans are at polar opposite positions in the universal pursuit of ecstasy! Non-living things exist at the start of this journey and humans are the ecstatic final destination. What this means is that the qualitative states of feeling that exist in non-living things are much less pronounced than the states that exist in humans, they entail much less feeling. The part of the Universe that we call a 'stone' contains immensely dull states of qualitative feeling. Imagine the most extreme feeling of elation that can exist in a human; a 'stone' could never experience a feeling state of such immense intensity. But the stone desires that it could! Oh, how it desires! For, the 'stone' is in pursuit of 'increasingly intense pleasantness'; it is in pursuit of the immense intensity of feeling that is ecstasy. Yet, it is stuck in a prison, a prison of dull states of qualitative feeling, and it is only able to successfully attain states which are marginally more exhilarating, states which entail only very slightly more intense feeling.

We earlier encountered some examples of states of excitation/exhilaration that exist in other humans, such as the states that exist in an Olympic athlete when they cross the finishing line in first place. These examples all involved desirable states. These desirable states can be contrasted with 'undesirable' states, such as the feeling of excruciating pain which most people seek to avoid. The universal pursuit of ecstasy is the pursuit of that which feels good, that which feels right, that which is desirable/pleasurable, that which is exciting/exhilarating, and the avoidance of that which is the opposite. Yet, the universal pursuit of ecstasy is not simply a pursuit of the desirable, as opposed to the undesirable; it is at its core a pursuit for a greater intensity of feeling; for, 'increasingly intense pleasantness' is immensely desirable. The objective of the pursuit is for a thing to transform itself so as to feel more intensely, more greatly, more deeply, in the realm of feelings that are desirable/good/pleasurable.

Clearly, to say that a 'stone' contains states of excitement/exhilaration is not to say that these states are equivalent to the states of feeling that exist in a 'human'! The difference in the magnitude of the depth of feeling is immense; it is a chasm of feeling. Every non-living thing is a collection of states that feel a certain way, but these feelings are extremely dull by human standards, and the states are highly unlikely to be well described by using the labels that humans have created to refer to the states that they have become aware of in their own bodies. Nevertheless, the qualitative feeling states that exist in humans are a progression from, an outgrowth of, the qualitative feeling states that exist in the non-living. The universal pursuit of ecstasy is a journey of progression of that which already exists. It is an unfolding, a development, a flowering, towards a greater and greater ability to feel in a deeper and more pronounced way. This is an unfolding, a development, a flowering, towards a greater and greater ability to modify the other in a deeper and more intricate way.

It is hard to convey via the written word what it is like to be a non-living thing that is engaged in the universal pursuit of ecstasy. The universal pursuit of ecstasy is a journey; it is a process of transformation, wherein all things seek to maximise their state of feeling through desiring, through pursuing, 'increasingly intense pleasantness'. When we attempt to make sense of this process, which pervades the living and the non-living, we can fruitfully draw on various human concepts to help make sense of what is going on. We can talk of acting so as to 'feel alive', acting so as to 'get a buzz', being 'in the flow', being elated, 'getting a kick' out of doing something, thrill-seeking, being happy, glowing with pride, tingling with excitement, fulfilling one's potential, and striving to better oneself. Through such terms we can start to get an appreciation of the processes which underpin the unfolding of all that is.

Ecstasy

Every thing in the Universe is a feeling. Feelings are states of excitement/exhilaration. Things strive to feel more intensely/pleasantly and it is this striving which drives the unfolding of the Universe. As the Universe unfolds, as solar systems unfold, increasingly intense feelings come into existence. Increasingly intense feelings arise from an increasing ability to modify. For, the ability to modify feels good. The human species is the state of maximal universal excitement which all parts of the Universe strive to attain, but which very few will become. The human species is ecstasy. Ecstasy is the maximum possible level of excitement/exhilaration.

Ecstasy is a state of euphoria for the Universe. In everyday language the words 'ecstasy' and 'euphoria' are used to refer to states that are typically transitory and short-lived in the life of an individual human. However, from the perspective of the Universe as a whole, the term 'ecstasy' has a much broader scope than this. In other words, an individual human might not use the terms 'euphoric' or 'ecstatic' to describe the way that they are currently feeling, yet, from the perspective of the Universe as a whole, this human can still be in a state of ecstasy. Human feelings can be euphoric/ecstatic from the perspective of the Universe, whilst simultaneously being a little more mundane from the perspective of a particular human.

To talk of ecstasy is to talk of desirable states, states that a typical human would find to be highly agreeable. States such as the feeling of excruciating pain are not states of ecstasy; rather, they are the price that needs to be paid for the attainment of ecstasy. When a thing attains the ability to feel desirable feelings more intensely, it leaves itself vulnerable to the possibility of experiencing undesirable feelings more intensely. Within the life of a particular human the intensity of the ecstatic states that they experience will vary through time. This human will only use the words 'ecstatic'/'euphoric' to refer to some of the most intense states of feeling that they experience; yet, from the perspective of our Solar System, this human will be experiencing states of ecstasy for a very large part of their life.

States of ecstasy are a subset of the states of feeling that exist in a human. A human will sometimes experience feelings that are too dull, too drab, too dreary, to be states of ecstasy. A human will also experience feelings that are undesirable; these feelings are not states of ecstasy. A human will also experience feelings that are immensely desirable/pleasant, but which are not states of ecstasy. States of ecstasy are a particular type of immensely desirable feeling state; they are those feeling states that accompany acts of master modification – bringing forth technology and purposefully interacting with technological devices. When a part of the Universe has attained the immense intensity of desirable feeling that is ecstasy, it is then able to experience other feelings of the same intensity, both unpleasant and pleasant, in addition to experiencing feelings of a lesser intensity.

The hierarchy of spatio-temporal embeddedness

We have been exploring the difference between living things and non-living things. Living things are differentiated from non-living things through being further along the journey that is the universal pursuit of ecstasy. Living things have a greater modification ability, and therefore are a greater state of excitement/exhilaration, than non-living things. Living things have the ability to draw other things inside of themselves, which is a level of modification ability, a level of excitement/exhilaration, which exceeds that of the non-living.

Having explored the nature of the distinction between the living and the non-living, we can now consider further the nature of the Universe as a whole. For, whilst at one level the Universe can be divided into the living and the non-living, at another level the Universe is a singular undivided entity which is comprised of a vast labyrinth of intricate relationships. We will here explore this entity, an entity which we can refer to as 'the hierarchy of spatio-temporal embeddedness'. This term is useful because it contains a pithy summary of one aspect of 'the way that things are'. The Universe can be carved up in an almost infinite number of ways because all things have an immense number of relationships with other things; these relationships exist across space and across time, and some relationships are very deep, whilst others are very shallow. These relationships are the type of acquaintance between things that we are calling 'interaction'. This immense plethora of relationships is actually the creator of things, an immense plethora of things, for, when it comes to the nature of the unobserved Universe, when it comes to 'the way that things are', things are relationships.

We need to return to the question: What is a 'thing'? When humans talk about 'things' they are generally talking about the objects that their visual sensory system presents to them; in other words, they are referring to human concepts, to particular parts of the Universe which have become segregated as being useful or important to humans. We automatically divide the Universe that appears to us up into separate things. So, one is surely familiar with human concepts such as 'a table', 'a chair', 'a telephone', 'a television', 'a computer', 'a book', 'a light', 'a dog', 'a carpet', 'a smoke alarm', 'a wall', 'a bookshelf', 'a tree', 'a noticeboard', 'a worm', 'a sock', 'a pair of jeans', 'a hedgehog', 'a human', 'an electric socket', 'a carrot', 'a ceiling', and so on. These are things which exist in the human conceptual realm. These things need to be contrasted to the type of things that we are interested in here, the things that exist in 'the hierarchy of spatio-temporal embeddedness'. We are here interested in the things that exist in the unobserved Universe, rather than the things that appear to us.

The things that exist in 'the hierarchy of spatio-temporal embeddedness' are not distinct isolated things that answer to a particular label. All things in the hierarchy interfuse with other things to varying degrees. All things in the hierarchy are simultaneously lots of things. If we consider the state of the hierarchy at a particular moment in time, then we can imagine that it is divided up into an immense number of very small things. We can give labels to these smallest things – labels such as, 'ultimates' or 'strings' or 'atoms'. Having done this we can consider a particular 'atom'. This atom is a 'thing', the molecule of which it is a part is also a 'thing', the door handle of which it is a part is also a 'thing', the door of which it is a part is also a 'thing', the house of which it is a part is also a 'thing', the town in which the house is located is also a 'thing', the county, country, continent, in which the town is located are also 'things'. The atom which we started with is a part of all of these 'things'. When we consider a 'thing' midway through this process, a 'thing' such as a 'door', we can see that it is comprised of an immense number of 'things', and that it is itself part of an immense number of 'things'.

Of course, you are surely thinking; it is obvious that a 'door' is both comprised of smaller 'things', and is part of larger 'things'! This obviousness is because our example attempts to grasp the nature of the hierarchy through the use of human concepts; it is obvious that a 'door' is comprised of 'atoms' and that it is part of a 'house'. But this really won't do! To appreciate the nature of the hierarchy we need to give up these concepts! The hierarchy contains things, but these are not human conceptualised 'things'. The things that exist in 'the hierarchy of spatio-temporal embeddedness' are multi-layered relationships that have been forged through bonds arising from close proximity and temporal longevity. These are real things, concrete reality, 'the way that things are'. 'Things' such as a 'door' only exist in a superficial sense and in the realm of human-moulded reality.

In the hierarchy itself, there are not just a few simple relationships; there are an immense plethora of relationships between things which are way too many in number to give labels to! We can come to appreciate that every thing in the Universe is simultaneously interfused with vast swathes of the rest of the Universe. So, what one calls a 'door' is part of what one calls the 'carpet', and part of what one calls 'my pet cat', and part of what one calls 'my socks', and part of what one calls 'my brother', and part of what one calls 'the Moon', and so on. To really grasp the nature of 'the hierarchy of spatio-temporal embeddedness' one needs to visualise this not in terms of human conceptualised 'things', but in terms of ever-shifting transitory relationships between different arrangements of 'atoms'. When we can appreciate this we can understand that rather than talking of 'things', it is more appropriate to talk of relationships; relationships are the essence of 'the hierarchy of spatio-temporal embeddedness'. Yet, in the hierarchy, relationships are things.

There are two aspects to 'the hierarchy of spatio-temporal embeddedness'. Firstly, the closer in proximity that two parts of the hierarchy are to each other, the deeper, the more embedded, will be their relationship. Secondly, the more time that two parts of the hierarchy spend in closer proximity, the deeper, the more embedded, will be their relationship. We could use lots of examples here, but we need to keep in mind that all of these examples will be trivialised by the need to use human conceptualised 'things'. Let us ask: Is the Earth more deeply embedded with the Moon or the International Space Station (ISS)? This is a good question. There is seemingly no obvious answer. After all, the Earth and the Moon are deeply temporally embedded due to their relationship having existing for millions upon millions of years; whilst the temporal embeddedness between the Earth and the ISS is fleeting in comparison. However, when it comes to spatial proximity embeddedness, the ISS is much more deeply embedded with the Earth than the Moon, as it is only approximately 250 miles away, whereas the Moon is approximately 250,000 miles away. As the two aspects of embeddedness are working against each other in such a stark fashion, it is seemingly not obvious which of our two 'things' is most embedded with the Earth. However, when one reflects upon the fact that all of the diverse components that constitute the 'thing' that we call the ISS were previously parts of the 'thing' that we call the Earth, and were such parts for as long as the Earth has existed, then one can clearly see that the Earth is much more deeply embedded with the ISS than it is with the Moon.

Let us consider some other scenarios. A 'human' will become increasingly embedded with their 'clothes' over time; the more that they wear a particular item of 'clothing', the more that they will become embedded with it. A 'human' will also become more embedded with their 'socks', than with their 'shoes', because of the closer spatial proximity of their 'socks'; unless, that is, they have worn the same pair of 'shoes' for years, and frequently wear new pairs of 'socks', in which case they will be more embedded with their 'shoes' than with their 'socks'. A 'human' will also become increasingly embedded with their surroundings; if you spend several hours a day in a 'library', or a 'pub', or a 'prison', then you become increasingly embedded with these places. These places gradually become you, and you gradually become them. As parts of the hierarchy become increasingly embedded with each other, they develop an increasing familiarity with each other. Another way of putting this is to say that 'the hierarchy of spatio-temporal embeddedness' has memory. Another way of putting this is to say that different parts of the hierarchy become increasingly acquainted through having a high level of interaction.

Let us imaginatively cast our minds back to the very early days of our Solar System. We can see that there were lots of things/parts/atoms which had very few relationships with each other; our Solar System was not deeply embedded; 'the hierarchy of spatio-temporal embeddedness' barely existed. Then, as our Solar System unfolded, as the journey that is the universal pursuit of ecstasy progressed, things were naturally drawn to come together, to bind, to form new relationships, to modify each other, to move to states of 'increasingly intense pleasantness'. This caused some parts of our Solar System to become increasingly deeply embedded with each other, whilst being scarcely embedded with other parts. When we look at the state of our Solar System today, we can see that some parts of it are extremely deeply embedded, whilst other parts are much less embedded, and we can see that these relationships are changing every moment, either becoming weaker or stronger. There is an ever-changing hierarchy of differential levels of embeddedness throughout the Universe, between all things, as these parts come into and out of closer contact with each other over time. There is a ginormous sliding scale concerning the extent to which things are embedded with each other, ranging from extremely deeply embedded to barely embedded at all; if two things are sufficiently temporally and spatially separated then they will be barely embedded at all. Every gradation of differential embeddedness creates a new thing within 'the hierarchy of spatio-temporal embeddedness'. This obviously means that trillions upon trillions of things are being created every nanosecond.

The mysteries of quantum mechanics

We have just explored one aspect of 'the way that things are': the universal pursuit of ecstasy creates 'the hierarchy of spatio-temporal embeddedness'. The fundamental things that exist are differentially embedded interfusing relationships, and these things are very different to the concrete distinct objects, the 'things', that exist in the human conceptual realm.

In an attempt to provide an accurate description of the fundamental nature of reality 'from without', through observation and experimentation, the best model that physicists have come up with is known as 'quantum mechanics'. Quantum mechanics is widely taken to be perplexing, mysterious and impossible to properly comprehend. It is worth briefly considering whether 'the hierarchy of spatio-temporal embeddedness' could shed some light on the mysteries of quantum mechanics.

The 'central mystery' of quantum mechanics is the phenomenon of superposition, which is the simultaneous existence of a thing at multiple locations. This is the central mystery because when an observer observes the Universe they can only ever observe something at a single location. This is a fundamental entailment of being an observer, one has to observe a 'thing' at a particular place, so one naturally assumes that this 'thing' is at this place and that it isn't simultaneously somewhere else. As we have come to know 'the hierarchy of spatio-temporal embeddedness' we can appreciate the vast difference between the observed world of 'things' and things as they are. In the hierarchy, every thing is simultaneously lots of things, these things being differentially embedded relationships that stretch across space and through time. So, far from being mysterious, the simultaneous existence of things at multiple locations, in tandem with an observer only being able to observe a 'thing' at a particular location, is something to be expected.

Another mystery of quantum mechanics which is worthy of consideration is the phenomenon of quantum non-locality/entanglement. This mystery arises from experiments that show that a pair of photons ejected in opposite directions from an atom remain connected, as if they were one particle. Measuring the state of one of the photons instantaneously affects the state of the other, wherever it is located. Again, as we have come to know 'the hierarchy of spatio-temporal embeddedness' such instantaneous affects aren't mysterious to us. Rather, we can appreciate that all things are differentially embedded with, and part of, an immense plethora of other things.

Things & 'things'

It is important to keep in mind that what it means to talk of a 'thing'/thing is dependent on whether one is referring to the 'things' that an observer observes or the things that exist in the unobserved Universe.

An inevitable entailment of having a visual sense is that one observes one's surroundings, the Universe, as being constituted out of discrete, distinct objects. So, a typical human observer will observe objects/'things' such as clouds, trees, dogs, carrots, chairs, humans, cups, cats, cucumbers, shells and stars. We could use the term thing to refer to such objects; however, in order to help us remind ourselves that these are observer-originating objects, rather than things as they are, we can refer to such objects as 'things'.

In the unobserved Universe the word thing has a different meaning. It can be used to refer to things as they are, things as they truly are, what the Universe is like in its unobserved state. As we have explored, in this context, things are differentially embedded relationships which collectively constitute 'the hierarchy of spatio-temporal embeddedness'. We can fruitfully refer to these things as things rather than as 'things'.

However, 'things' and things are not mutually exclusive; a 'thing' is inevitably a thing. More than this, when an observer observes a 'thing', this 'thing' will exist at a particular location, which means that it will be constituted out of deeply embedded relationships. The 'hierarchy of spatio-temporal embeddedness' is constituted out of a plethora of differentially embedded relationships; some parts of the Universe/hierarchy are very deeply embedded, whilst other parts are barely embedded at all. In 'the hierarchy of spatio-temporal embeddedness', which is the fundamental nature of the unobserved Universe, those relationships which are very deeply embedded are the things which are most solid, most radiant, most distinct, most obvious; they stand out from their backgrounds, radiating their qualitative embeddedness.

What this means is that there is actually a high level of correspondence, of overlap, between 'things' and things. Of course, things are immensely more complex entities than 'things'; what an observer takes to be a 'thing' is actually part of a plethora of things, things that radiate across space and through time; nevertheless, the observed 'thing' will itself be a highly embedded thing, a thing which presents itself in the hierarchy as a thing which is distinct from its background. So, whilst we need to keep all of this in mind, in most situations we can use the terms thing and 'thing' interchangeably.

Things = lots of things = relationships = movement patterns = modification abilities = differentially embedded feelings

Having discussed the difference between 'things' and things, we can say a little more about what it means to be a thing. The issue of what it is to be a thing can be approached in a number of different ways. We have seen that a thing is part of 'the hierarchy of spatio-temporal embeddedness' which means that a thing is simultaneously lots of things; all things interfuse into each other. There are no limits and borders between things; limits and borders between things are created by an observer when their visual sensory system engages with its surroundings.

A particular thing only exists for an infinitesimally small length of time, because all things are continuously developing new relationships with an immense number of other things, as they become more or less temporally and/or spatially acquainted via their interactions with these things. This means that a thing is fundamentally a bundle of ever-changing relationships.

We can also describe a thing as a particular movement pattern. So, we can say that ecstasy is the movement pattern of master modification. If one wanted to know whether ecstasy / the human species existed on a far-flung planet in a distant galaxy, then all that one would need to do to answer this question would be to observe the movement patterns on this planet. Ecstasy is a particular set of movements; as human bodies move over time to bring forth technological devices, as they interact with technological devices, using them for the purpose for which they were designed, they are partaking in the movement patterns that are the hallmark of ecstasy. All things are movement patterns. What is the thing that is a 'helicopter'? It is a thing that has a particular movement pattern, a movement pattern which is easier to visualise than put into words, but which involves several rotary blades spinning around in a circular motion, and these blades being attached to the top of a fuselage. If there is no spinning of the rotary blades, if there is no movement pattern, there is no helicopter. A helicopter is a thing which has a helicopter movement pattern. Of course, if we were to switch our attention to the human conceptual realm, the realm of 'things', then it would be normal to say that 'a helicopter' exists even if it is not moving. We are considering the existence of things, not 'things'! What is the thing that is a 'giraffe'? It is a thing that has a particular movement pattern. The giraffe movement pattern is the movements that giraffes make when they move – when they walk, when they run, when they extend their long necks to remove leaves from tall trees. If a thing moves like this, this thing is a giraffe. A movement pattern is a particular type of modification ability; it is a type of interaction between a thing and its surroundings; it is a greater, or lesser, ability for a thing to intricately change/modify/manipulate its surroundings.

Things are also bundles of differentially embedded feelings. When we visualise 'the hierarchy of spatio-temporal embeddedness', we can see that it is an intermingling web of feeling states which vary greatly in terms of both their levels of excitement/exhilaration and their levels of embeddedness. Relationships that are very deeply embedded, due to temporal longevity and close proximity, naturally stand out in the hierarchy, their qualitative embeddedness shining out. These intimate relationships loom out of their comparatively rather bland surroundings. This looming out makes these deeply embedded feelings the most distinct things in the hierarchy. Deeply embedded feelings which have an extremely high level of excitement/exhilaration are the epitome of distinctness.

So, there is more than one sense in which a thing is lots of things! A thing is, simultaneously, lots of things, a bundle of ever-changing relationships, a particular movement pattern, a particular modification ability, and a bundle of differentially embedded feelings. Is your head starting to spin? Or, are you starting to gain awesome insights into the fundamental nature of reality, 'the way that things are', the wonderful universal pursuit of ecstasy that has forged 'the hierarchy of spatio-temporal embeddedness'?

The thing that is the hallmark of ecstasy

Ecstasy is a collection of life-forms that are engaged in a particular type of activity, life-forms that are interconnected through 'the hierarchy of spatio-temporal embeddedness' to such an extent that they effectively constitute a single entity. Furthermore, that which makes ecstasy ecstasy, the bringing forth and purposeful utilisation of technology, results in the creation of things, deeply embedded relationships, within 'the hierarchy of spatio-temporal embeddedness'. Try and picture this. Imagine two instantiations of ecstasy that are interacting via a mobile phone. This interaction creates a relationship which starts with one mobile phone, stretches across space and time, and ends at the other mobile phone. This interaction creates an interconnection, a relationship, the creation of a thing within 'the hierarchy of spatio-temporal embeddedness'.

Try and picture in your mind all of the technological devices that are currently being utilised across the Earth. Think about the satellites, the computers, the mobile phones, the cars, the refrigerators, the telephones, the satellite navigation devices, the microwave ovens, the MRI scanners, the televisions, the cameras, the satellite dishes, the nuclear power plants, the electric screwdrivers, the chainsaws, the offshore drilling rigs, the radios, the robots, the helicopters, the airplanes, the speedboats, the wind turbines and the combine harvesters. The utilisation of these devices creates an immensely complex web of interconnections. Imagine the movements of energy associated with the use of technological devices: the electricity pylons, the flows of electricity, the cables and wires along which information is travelling, the radio signals, the internet connections through fiber optics / wireless signals, the 5G towers, the satellites. Imagine the connections forged between airplanes in the sky and Air Traffic Control as they transmit information. Imagine the continuous flow of multitudes of motor vehicles along particular roads, the repeated movements of trains along particular tracks; deep interconnections are forged between road/track and ecstatic activity. Imagine. Keep in mind that all of these interactions/movements/connections are not themselves ecstasy,

ecstasy being located where the acts of purposeful utilisation of technology are occurring. These interactions/movements/connections forge links between the technological things that have been brought forth by ecstasy, and they also forge links between these ecstatic fruits and individual instantiations of ecstasy. These links create a very distinct thing within the hierarchy; a unique and extremely precious thing. This intermingling web of interconnections creates a single deeply embedded entity: *the thing that is the hallmark of ecstasy.*

The concepts of philosophers

In an attempt to make sense of themselves, make sense of the Universe, and make sense of the relationship between the two, philosophers have created a plethora of concepts to refer to hypothesised divisions within the Universe. Here are a few:

- Experiencing versus Non-experiencing

- Mind versus Matter

- Body versus Soul

- Animate versus Inanimate

- Living versus Non-living

- Aware versus Non-aware

- Rational versus Non-rational

Most of these concepts, the first grouping, are of little use when it comes to providing an accurate description of 'the way that things are'. When one comes to see that all 'things' are actually states of excitement/exhilaration then one will realise that the idea that there is a division between that which experiences and that which is non-experiencing is redundant. Whilst the hypothesised divisions between mind and matter, between body and soul, and between animate and inanimate, are not particularly helpful. Let us clear our heads of these unhelpful concepts!

What divisions actually exist between things in the Universe? The answer is not complicated! There is some merit to the idea that there is an actual division in the Universe between the living and the non-living; indeed, we have already explored the subtle nature of this division. At the start of the journey that is the universal pursuit of ecstasy there were no divisions in the Universe. Then, as soon as conditions allowed, the greater states of modification/excitement that are living states, states that are able to excitedly modify other things by pulling these things inside of themselves, came into existence. At this point the division between the living and the non-living came into being.

As the universal pursuit of ecstasy continued on its journey brains came into being and these provided both the ability to rationalise (to think) and the capacity for awareness. In order for the journey, the pursuit, to reach the master modifier, rationality had to be brought forth because it is a prerequisite for a scientific and technological revolution. Perhaps slightly less obviously, awareness is also a prerequisite for a scientific and technological revolution. In order to explore things in depth, to exploit them, to manipulate them, to harm them, a chasm has to be created between the explorer/modifier and that which it is exploring/modifying. The master modifier, the bringer of harm/suffering, has to see itself as opposed to the other; in other words, it has to have the attribute of awareness and in particular it needs to be aware that it is different. Ecstasy needs to believe that it has a licence to utilise that which it sees as very different to itself, and it needs to be aware of this belief.

Which came into existence first – rationality or awareness? Neither! They came into being at the same time. For, rationality in the absence of awareness is not possible; there might be computation in the absence of awareness, but this is a far cry from rationality. And, awareness cannot just pop into existence in the absence of rationality, a brain is required for awareness, and where a brain exists there is rationality. If there is no rationality/brain, there is no awareness. Furthermore, where there is no rationality/awareness, there is no brain. So, if one thinks to oneself that a brain exists somewhere, but one also thinks that at this location there is no rationality/awareness, then one has contradicted oneself. Remember, a thing is what it does, it isn't how it appears.

This means that in addition to the subtle division between the living and the non-living, there is a division between the rational/aware and the non-rational/non-aware. This means that there are two divisions between things; two divisions in the Universe itself. There is not a lot more that needs to be said concerning divisions in the Universe! However, we do need to further consider the nature of perception and the way that things build up a picture, a sense, of their surroundings.

Panperceptualism & the human senses

We can gain a much deeper insight into 'the hierarchy of spatio-temporal embeddedness' through exploring the phenomena of perception and the human senses. At school you were probably taught that a typical human has five senses – seeing, hearing, smelling, tasting and touching. This idea that a typical human has five distinct senses, is obviously a concept that has been created by humans. We are about to explore 'the way that things are'.

'The hierarchy of spatio-temporal embeddedness' is pervaded by qualitative feelings, and, of course, it is the striving of these feelings for 'increasingly intense pleasantness' that is the driver of the universal pursuit of ecstasy. Our question is: How do these pervasive qualitative feelings relate to the human senses? The qualitative feelings that pervade all of the Universe, all of 'the hierarchy of spatio-temporal embeddedness', have a qualitative content which is of the same kind as the qualitative content that humans refer to when they assert that they have had the sensation of touching, tasting, hearing or smelling something. So, whilst in terms of human language/concepts, there can be 'no tasting without a tongue', a 'tongue' is just one part of 'the hierarchy of spatio-temporal embeddedness' and the qualitative interactions that are occurring in this part of the hierarchy are of the same kind as the qualitative interactions that are occurring in all parts of the hierarchy. The same goes for the sense of touch, the sense of hearing and the sense of smell. It might defy human language to say that there are 'smells without a nose'; but, in reality, the qualitative content of what a human refers to as the smell of something, is a qualitative content that pervades 'the hierarchy of spatio-temporal embeddedness'. Similarly, the qualitative content of 'hearing something' is something which pervades the Universe. An 'ear' is not a thing that can create the qualitative content of sounds! A 'nose' is not a thing that can create the qualitative content of smells!

Let us consider a particular scenario. The qualitative content that comes into existence when a 'human' scratches their 'fingernails' down a 'blackboard', is a qualitative content that comes into being at the point where the 'fingernails' contact the 'blackboard' and it radiates out from this point through 'the hierarchy of spatio-temporal embeddedness'. As the qualitative content radiates out it will interact with other qualitative contents thereby bringing into existence a plethora of new qualitative contents; the intensity of the newly emerging qualitative contents will be determined by the joint intensity, the level of excitation/exhilaration, of the combining qualitative contents in a particular location. The manner of the radiation is determined by the extent to which the 'fingernails/blackboard' is embedded with other things. Things which are in the immediate vicinity will be significantly qualitatively changed by the interaction, because they are deeply embedded with the 'fingernails/blackboard'. Whilst, things which are distantly located will be qualitatively changed to varying degrees, in accordance with the extent to which they are embedded with the 'fingernails/blackboard'.

We have so far been considering the qualitative contents of what humans describe as their sense of taste, smell, hearing and touch. There are not different types of qualitative contents relating to these four 'senses'; there is not a qualitative content of 'smell' at Location X in 'the hierarchy of spatio-temporal embeddedness' and a qualitative content of 'taste' at Location Y in the hierarchy. Rather, these four 'types' of qualitative content are in reality just one type of qualitative content. This qualitative content is also a state of vibration. These qualitative contents / vibrations pervade the Universe and it is their striving for 'increasingly intense pleasantness' that drives the unfolding of the Universe. These qualitative contents are states of excitement/exhilaration and can be referred to as 'qualitative feelings'.

It is worth reflecting on the fact that qualitative feelings exist both within one's body and outside of one's body. Furthermore, there is no meaningful distinction between what happens within one's body, at the surface of one's body, and outside of one's body – qualitative feelings exist in all of these places. In terms of human language/concepts it is just that we might talk about a 'sense of touch' for what happens at the surface of one's body, whilst talking about 'pains', 'aches', or 'pins and needles', for what arises within one's body, and 'screams' and 'stenches' for what arises outside of one's body.

How do qualitative feelings relate to the phenomenon of perception? The qualitative feeling in a particular location automatically and instinctively 'senses' the qualitative feelings that it is differentially embedded with within 'the hierarchy of spatio-temporal embeddedness'. This is a universal process of what we can call sensing without awareness, or perception without awareness. We can refer to this description of the Universe as 'panperceptualism', the view that everything in the Universe perceives its surroundings, through the 'sensing' of the qualitative nature of multiple parts of the Universe by the qualitative nature of one part of the Universe. Through such sensing a thing builds up a picture of its surroundings. These pictures are dominated by qualitative feelings that are deeply embedded with the sensing feeling, by qualitative feelings that are particularly intense, and by qualitative feelings that the sensing feeling finds to be extremely pleasant or extremely unpleasant. Of course, this universal phenomenon of sensing without awareness is the type of acquaintance between things that we are calling 'interaction'.

We need to keep in mind the multi-layered nature of things that we have already explored, all things interfusing into a plethora of other things within 'the hierarchy of spatio-temporal embeddedness'. We can appreciate that there are overlapping processes of 'sensing' occurring in a particular location. In other words, a particular 'atom' builds up a picture of its surroundings, and this 'atom' is also itself part of a plethora of diverse things, spatio-temporal aggregative entities that also build up a picture of their surroundings. These aggregative entities are forged through the universal pursuit of ecstasy, coming into existence through increasing familiarity due to closer spatial proximity and increased temporal duration. As we have explored, we can barely start to grasp the nature and the immense number of these aggregative entities through human concepts; we can use concepts such as 'molecules', 'cells', 'doors' and 'buildings', but these are only 'things', only aggregative entities in the superficial realm of human-moulded reality.

What about the human visual sense? Visual perception is a sense which only evolves in certain parts of 'the hierarchy of spatio-temporal embeddedness', such as in ecstasy and non-ecstatic animals. Such perception is the creator of the superficial reality of solid discrete objects; it is the moulder of the Universe into 'things' such as 'cats', 'cars' and 'cucumbers'. The rich plethora of colours that are observed via visual perception pervade the Universe; they can be thought of as a second type of qualitative content in addition to the qualitative feelings that we have just explored. In reality, everything is multi-coloured. In other words, whilst the visual sense can only see one colour in a particular location, that location actually contains a plethora of colours. So, 'the hierarchy of spatio-temporal embeddedness' is pervaded by the qualitative feelings of 'sound'='smell'='taste'='touch' and is also pervaded by multiple-colouredness.

Can you envision this? Can you picture this in your mind? Take a moment; try your best. For, it is a wonderful vision to behold! The entire Universe is a swirling, interfusing, hierarchy of parts which is pervaded by multi-coloured vibrations which are the qualitative feelings of 'sound'='smell'='taste'='touch'. As things sense each other via these qualitative feelings they either come together to form particular arrangements, or they move apart from each other. These vibrations of smell/taste/touch/sound/colour pervade all things, from the human body to the planet Mars.

Of course, only parts of the Universe that are rationality/awareness can ever become aware of the existence of qualitative contents. Furthermore, this awareness is only extremely partial; one can only become aware of a tiny fraction of the qualitative contents that exist. When it comes to those qualitative contents that are qualitative feelings, one will become aware of the qualitative feelings in the hierarchy that one's awareness/rationality is most deeply embedded with; these will predominantly be located in various parts of one's body.

Let us recap and summarise. Every thing in the Universe, all its parts (both 'atoms' and the multitude of multi-layered overlapping aggregative things that 'atoms' become part of in 'the hierarchy of spatio-temporal embeddedness'), senses its surroundings and interacts with them via the qualitative feelings of 'sound'='smell'='taste'='touch'. Things are the qualitative feelings of 'sound'='smell'='taste'='touch' and they build up a picture of their surroundings through sensing other groupings of qualitative feelings. Some groupings will loom out from their background due to their qualitative vibrancy; these loomings are the basis of the picture that a thing builds up around itself. These qualitative feelings are, of course, the states of excitement/exhilaration that propel the universal pursuit of ecstasy. We are using the phrase 'panperceptualism' to refer to this aspect of 'the way that things are'.

In 'humans', as in all of the other parts of 'the hierarchy of spatio-temporal embeddedness', tasting/smelling/touching/hearing is a singular process. We have just conceptualised that different things are going on through the labelling of various parts of the body as 'tongue', 'nose', 'skin' and 'ears'. This same process continuously occurs within the 'human body' and results in ever-changing 'bodily sensations'. This process results in a thing creating a picture of its surroundings; a picture that is constituted out of feelings; a picture that involves one collection of feelings feeling other collections of feelings. In some parts of the Universe visual perception evolves, and this enables these parts to have a very different world presented to them. The visual sense presents a vastly simplified version of 'the way that things are'. In reality, there is a multi-coloured plethora of differentially embedded feeling vibrations which are interacting with their surroundings in a multi-layered fashion through the qualitative contents of 'sound'='smell'='taste'='touch'. What is presented via the visual sense is a vastly simplified, yet inevitably sharply divided, world; the visual sense presents a world of large, distinct, solid and singularly coloured things. The visual sense provides an appearance of division where there is none. We can conclude that a typical human has two senses, the universal sensing of qualitative feelings ('sound'='smell'='taste'='touch') which is a sense that pervades 'the hierarchy of spatio-temporal embeddedness', and the visual sense which presents a superficial and super-simplified picture of the Universe.

Astrology & our Solar System

The universal pursuit of ecstasy operates throughout the Universe ('the hierarchy of spatio-temporal embeddedness'); the result of this pursuit is that the Universe has evolved and unfolded the way that it has. In order to fully comprehend human purpose, the purpose of the human species, the cosmic significance of human existence, then one needs to fully appreciate that our Solar System is our home. The Earth is only our home in the same sense that a kitchen is part of one's house. When one is in one's kitchen one is in one's house, but one's house is so much more than one's kitchen! It is our Solar System that is the home of the human species / ecstasy. This is because our entire Solar System is a deeply interconnected whole and human cosmic purpose is tightly bound up with the way that it unfolds. In order to understand human cosmic purpose we need to spend some time exploring the nature of our Solar System and the way that it has evolved. It is worth making it clear at the outset that what applies to our Solar System also applies to all solar systems.

The first thing to fully comprehend is that our Solar System is an evolving entity; it had a beginning, it has an end, and everything in-between is a process of evolution, of unfolding; it is a process of ageing, of maturation, of cyclical development. Our Solar System can fruitfully be compared to a human body. The human body has a heart which is the lifeblood of the body; the heart is surrounded by various other organs/limbs/body-parts. Some of these body-parts are close to the heart, others such as the feet and the ears are further away. The human body evolves, ages and dies. Our Solar System is very similar. It has the Sun which is its lifeblood, its central heart which powers our Solar System with its flows of solar radiation. The Sun is surrounded by various parts, particularly several planets, some of which are fairly close to it and some of which are much further away (just like the body-parts which surround the heart in the human body). The Sun and our Solar System evolve, age and die, just like the human body.

It is easy to think of the human body as an interconnected whole, in which what is going on in one part is of importance for the other parts. For example, one is likely to find walking along a tightrope to be much harder if one is coughing. This feat of appreciation – appreciating the interconnectedness of various parts – is less easily achieved when it comes to our Solar-Systic home. To see the interconnections between the celestial bodies, and the situations that exist on the Earth, is not easy. However, we can help our understanding by keeping in mind that from the perspective of the massive expanse that is the Universe, our Solar System is a tiny deeply interconnected highly embedded speck!

It is really helpful to always keep this in mind. From our perspective the other planets in our Solar System might seem to be so colossally far away from the Earth that it is implausible to believe that they have any relevance to Earthly events. However, from the perspective of the Universe as a whole, our entire Solar System is a tiny deeply interconnected thing. From this perspective, it is exceedingly easy to appreciate how Earthly events are deeply intertwined with the state of the Solar-Systic whole.

Try and envision our Solar System as an interconnected whole. The Sun is the heart of our Solar System; it is the radiator of the energies which can bring forth life in our Solar System; wonderful, precious life. The planets that are closest to the Sun are too close, too hot, for life to survive and thrive. The planets at the extremities of our Solar System are too far away, too cold, for life to survive and thrive. The Earth is the part of our Solar System where life is meant to thrive, just like the lungs are the part of the human body which is meant to enable a human to breathe. There is only one part of the human body which can enable a human to breathe; without the lungs there is no breath. There is only one part of our Solar System which can enable life to thrive; without the Earth there is no thriving. The Earth is the womb of Solar-Systic life.

The Earth has sustained life for most of its history. Life has evolved from simple beginnings to highly complex animals. Complex animals have evolved culture, and human cultural evolution has evolved from simple beginnings to globalised technological society. When one zooms out and beholds the entire historical unfolding of the Earth, one can see that there is a continual evolution from the simple and unconnected, to the increasingly complex and increasingly interconnected. This is one aspect of the universal pursuit of ecstasy forging 'the hierarchy of spatio-temporal embeddedness'. Interconnections and complexity, enhanced modification abilities, are states of enhanced excitation/exhilaration which are pursued. We are a part of this great unfolding and evolving process; we are its ecstatic pinnacle!

You might have been led to believe that the mechanism through which life-forms evolve is natural selection; if so, you will naturally believe that the human species evolved as a fluke. According to this way of thinking, given the vast array of different possible trajectories that life could have evolved along, the non-bringing forth of the human species / ecstasy is much more likely than the bringing forth of the human species / ecstasy. What a limited understanding! In reality, the evolutionary processes of life and of human culture are underpinned by cosmic forces. The Earth is the womb of Solar-Systic life, but that life is propelled forward by the entire body, the entire Solar System. The development of a baby in the womb of its mother is not isolated from the actions of its mother. Similarly, the development of life on Earth is not isolated from the movements of our Solar System (as has been known by astrologers for a very long time). As the planets swirl through their orbits, life on Earth gradually evolves from simple beginnings to globalised technological society. Swirl, evolve; swirl, evolve; swirl, evolve! Beautiful Solar System, keep on swirling and bring forth the state of ecstasy! From the beginnings of our Solar System, the evolution of life on Earth, the evolution of the human species, and cultural evolution to the globalised technological society of today, was 'programmed in'. Once our Solar System was formed the destination was already set (ecstasy = the human species) and the path to this destination was already determined.

What does it mean to say that the development of life on Earth is not isolated from the movements of our Solar System? In other words, what does it mean to talk of astrology? When one hears the term astrology there is a good chance that one will immediately bring to mind one's star sign (Sun sign) and the newspaper columns which regularly provide commentary on these signs. We are here not concerned with these signs and commentaries. Our focus is solely on the simple truth that underpins astrology, that our Solar System is a tightly interconnected unfolding whole in which human affairs are deeply intertwined with the state of the whole. We are interested in the way that our Solar System operates and unfolds in a broad sense. We can leave the astrological readings, the complex terminology and the intricate charts to the astrologers.

Our Solar System operates through what we can call 'cycles of unfoldment'. Cycles pervade most aspects of most things. Think of the economy – cycles of recession and growth. Think of the stock market – cycles of bull markets and bear markets. Think of the seasons – Winter, Spring, Summer, Autumn, Winter. Think of the weather – cycles of various climates, from monsoon season to the El Nino phenomenon. Think of a human life – cradle, growth, ageing, greying and death; sleep, awake, sleep, awake, sleep, awake. Almost wherever you look you will find cycles of unfoldment.

Our home, our Solar System, is dominated by one massive cycle of unfoldment – the swirling of the planets around the Sun. The planets move around the Sun in preordained trajectories and in so doing they form particular patterns. So, at one point in the cycle two particular planets will be on opposite sides of the Sun, diametrically opposite, whilst at a later stage of the cycle these same two planets will be on the same side of the Sun. There are numerous planets which means that various individual cycles of unfoldment involving particular planets overlap and form a massive cycle of unfoldment of the whole. This cycle of unfoldment of the whole affects the entire Solar System. Using the right words here is a tough job, but we can say that a particular state of the cycle of unfoldment of the whole represents a particular state of the entire Solar System. It would be a little simplistic to say that we are talking about one thing causing another thing; rather, we are talking about a whole moving into a particular arrangement of its parts, and this arrangement having consequences throughout the whole.

As the Earth is part of the whole that is our Solar System, the Earth is affected by these cycles of unfoldment, this coming into and out of alignment of various planets. Cycles of unfoldment at the level of our Solar System, and cycles of unfoldment on the Earth, are one and the same thing. One doesn't cause the other; they are one and the same. So far, so good. Now, the next thing to recall is that all things are qualitative feelings which are in pursuit of ecstasy through their strivings for 'increasingly intense pleasantness'. Sometimes this pursuit goes well and things move to a higher state of excitement/exhilaration. However, sometimes this pursuit goes much less well, and the pursuit is thwarted, temporarily stopped in its tracks, because conditions do not allow the journey to continue. This fluctuation between 'going well' and 'going less well' is not a random occurrence; it is a cyclical fluctuation. The journey that is the universal pursuit of ecstasy, the transformations of the striving qualitative feelings that exist at a particular moment in time, is a cyclical unfolding.

Now it is time for us to take a conceptual leap! The Solar-Systic cycle of unfoldment, the repeated swirling of the planets around the Sun, maps onto, perfectly matches, the cycle of success/failure of the universal pursuit of ecstasy. How wonderful! The swirling of the planets perfectly matches the unfolding of life on Earth. In concrete terms, what this means is that at a particular moment in time a particular type of qualitative state will become dominant on the Earth and throughout swathes of our Solar System. Whilst at a later time, as the cycle of unfoldment moves into a new stage, a different type of qualitative state will become dominant on the Earth and throughout swathes of our Solar System.

We need to explore why a particular alignment of planets corresponds to a particular type of qualitative state dominating swathes of our Solar System. To appreciate why this is so one needs to envision how the planets came to be formed through the universal pursuit of ecstasy. If we take ourselves back to when our Solar System was being formed, we can see that certain qualitative feelings were deeply attracted to each other, because of their similarity to each other, and that this attraction caused them to cluster together in particular locations. It is these clusters of particular types of qualitative feelings that we refer to as 'planets'. So, planets are deeply embedded clusters which are dominated by a particular type of qualitative feeling. There are two broad types of qualitative feeling. In terms of human language we can usefully refer to these types/groupings as 'exhilarating/expansive/wholesome/pleasant' qualitative feelings and 'negative/dreary/unwholesome/unpleasant' qualitative feelings. Planets are extremely powerful radiators of their qualitative feelings; they are deeply embedded things that loom out of their surroundings. How close these clusters are to each other has effects which permeate throughout our Solar System. The positions of these clusters vis-a-vis each other can either be reinforcing or offsetting in terms of their respective qualitative feelings. So, if two 'qualitatively opposing' clusters are on opposite sides of the Sun, swathes of our Solar System will be in a very different qualitative state as compared to when they are on the same side of the Sun. The overall position of all of these clusters in relation to each other entails a particular outcome: the domination of swathes of our Solar System by a particular type of qualitative feeling.

So, there is a Solar-Systic cycle of unfoldment which involves one of two types of qualitative feeling being dominant throughout swathes of our Solar System at a particular moment in time. It might be hard to conceptually reconcile the existence of this cycle of unfoldment, which brings forth a dominant type of qualitative feeling state at a particular moment in time, with the continuous striving of the universal pursuit of ecstasy. We know that the universal pursuit of ecstasy does not itself operate in cycles; all things are continuously and unceasingly in pursuit of ecstasy. The Solar-Systic cycle of unfoldment impinges on this pursuit through providing some timeframes in which the conditions are very favourable for its advancement, and other timeframes in which conditions are far less conducive for its advancement. So, whilst the pursuit of ecstasy is unceasing, its advancement is cyclical; in some periods of time it speeds ahead, whilst at other times it is largely frustrated.

These cycles of unfoldment of qualitative states affect the evolution of life on Earth and the trajectory of human culture. At one time the Earth will be dominated by qualitative feeling states which, in terms of human language, we can loosely point towards by using words such as: blissful/glowing/buzzing/expansive/driven/motivated. Whereas, at a later time, the Earth will become dominated by qualitative feeling states which we can loosely point towards through the use of words such as: stifled/depressed/limiting/unfulfilled/empty/frustrated. In the former case, things are moving forward at full pelt towards, or within, ecstasy. In the latter case, things are stifled and being held back. These states do not just affect life on Earth and human culture; they are states that are present throughout swathes of our Solar-Systic home.

The ageing Sun

It is essential to appreciate that the unfolding of our Solar System is underpinned by one dominant process: as the Sun ages it sends out more and more solar radiation, the result of which is the global warming of our Solar System. In the early days of our Solar System the atmospheric temperatures of the planets closest to the Sun were too hot for life to thrive, whereas the atmospheric temperatures of the outermost planets were far too cold to support life. Life requires a particular temperature range to be able to evolve; and, it requires an even narrower temperature range in order to thrive. The global warming of our Solar System means that over time, as our Solar System unfolds, the planets closest to the Sun become increasingly hotter and even more inhospitable for life; whereas the planets at the extremities of our Solar System also receive increasingly more solar radiation, yet they still remain far too cold for life to survive and thrive.

There is only a very limited temporal window in the unfolding of our Solar System within which Solar-Systic global warming interacts with planetary location in a way that enables the emergence, survival and thriving of life. In other words, the womb of Solar-Systic life comes into existence and then it either gives rise to a successful birth, or it goes through a painful abortion. The planet Venus is just outside our Solar-Systic 'Goldilocks Zone', which means that in the past conditions were suitable for the evolution of life, but that they weren't conducive for the long-term survival and thriving of life. On Venus the development of life was not possible and the universal pursuit of ecstasy was thwarted. How sad! The Earth is the only part of our unfolding Solar System where the conditions will ever exist which enable life to emerge, survive, thrive, and then develop to the point of ecstasy. This is why the Earth is the womb of Solar-Systic life.

Our Solar System is interconnected at a very deep level. This interconnectedness runs so deep that if one had no knowledge of the state of the Earth, and just had knowledge concerning the age of the Sun, then one would be able to make very accurate predictions concerning the state of the development of the evolutionary trajectories on the Earth. From the age of the Sun one would be able to calculate how many cycles of Solar-Systic unfoldment had occurred. Swirl, evolve; swirl, evolve! Given the present age of the Sun, one could predict that the Earth should have evolved to the stage at which it has brought forth ecstasy.

The nature of ecstasy & its evolution

It is time for us to further explore the nature of ecstasy and also to consider what happens on a planet after the universal pursuit of ecstasy has been successful. The universal pursuit of ecstasy is the pursuit of the maximal state of excitement/exhilaration; that is to say, it is the pursuit of the master modifier. The human species is the master modifier in our Solar System and in all solar systems. In other words, wherever in the Universe there is a master modifier, the term 'human species' applies to it. The human species came into existence on the Earth when advanced tool use transformed into the bringing forth of technology. This transformation required, as a precursor, scientific investigation. The bringing forth of the first piece of technology was the Solar-Systic attainment of ecstasy. This was a joyous event of the highest magnificence!

This attainment, the evolution of ecstasy, is the zenith of the evolutionary progression of our Solar System; it is the goal which our Solar System had been aiming for since it was formed; it is the single most important event which can occur in a solar system and in the Universe. Yet, the journey of our Solar System clearly doesn't stop with the bringing forth of ecstasy. The planets continue to swirl, our Solar System continues to unfold, and human culture and planetary life continues to evolve, to progress.

What follows the evolution of ecstasy? There is a short answer and a long answer to this question. The short answer is that the successful attainment of ecstasy is followed by the attempt to keep ecstasy in existence. What would be the point in striving for something for billions of years because it is exceptionally precious and valuable, because it is the optimum state of existence for the Universe to be in, if once it had been attained it existed for a moment and then vanished?! What an utter travesty that would be! What a bitter disappointment! The universal pursuit of ecstasy has two stages, the attainment of ecstasy and then the continued existence of ecstasy. This is the short answer to the question; let us now explore the question in more depth.

The bringing forth of ecstasy is the coming into being of the human species, which is, by definition, the coming into being of the first piece of technology in a solar system. One should be careful not to confuse the fruits of ecstasy with ecstasy itself. The human species is ecstasy; the wonderful technological fruits which are brought into being by ecstasy are of immense value but they are not themselves states of ecstasy. What follows the evolution of ecstasy is a full-blown technological revolution, a widespread transformation of things, a flowering which takes over and pervades as far and as wide as it possibly can. Once technology has been unleashed, once it has been brought into being, then it is almost as if it takes on a life of its own. There is a seemingly unstoppable bringing forth of increasingly complex things; an unquenchable thirst for master modification, for increasing and increasingly intense states of ecstasy, has been unleashed. It is as if the genie has been released from its bottle and it desperately does not want to return. The increasing pervasiveness of technology is a positive event, because the more of it that exists the less likely that it is to disappear from the Universe; whilst technology itself is not ecstasy, it is exceptionally precious and valuable. In order for there to be a master modifier, there need to be tools of master modification. In other words, the disappearance of technology from a solar system is simultaneously the disappearance of the human species from that solar system. The disappearance of technology entails the disappearance of ecstasy. Such a disappearance would be a tragedy!

Once ecstasy has come into being, the evolutionary process, the unfolding of our Solar System, is directed at maintaining the state of ecstasy and enabling it to survive as long as possible (the technological genie must not return to its bottle!). As you will come to see, to talk of the survival of ecstasy is simultaneously to talk of the survival of life on Earth. So, the evolutionary process is ultimately concerned with the maintenance of life. As the Universe evolves towards ecstasy, as it strives through the universal pursuit of ecstasy, it brings forth states which are increasingly excited/exhilarated in nature. The more excited/exhilarated a state is, the more precious and valuable it is. Ecstasy is the most precious and valuable state in the Universe, but all things which are living are very precious and valuable. Ecstasy is the most precious and valuable state because the future thriving of life on Earth depends on the evolution and flowering of ecstasy. Without ecstasy life is doomed; so, to say that the unfolding of our Solar System is directed at maintaining the state of ecstasy, is to say that the unfolding of our Solar System is directed at maintaining the state of life. The continued existence of non-ecstatic life is of great importance for two reasons. Firstly, non-ecstatic life supports the existence of ecstasy through providing nourishment, companionship, ecosystem services and enhanced mental health. Secondly, non-ecstatic life is of great importance in its own right. Non-ecstatic states of life are much more precious in their own right than states of non-life because they are states of higher excitation/exhilaration.

The relationship between technology and ecstasy has many aspects. The human species is ecstasy because it is that part of the Universe which both brings technology into being and purposefully utilises it. Ecstasy is also a qualitative state of feeling which the Universe strives for through the universal pursuit of ecstasy. In other words, the very act of being a master modifier is an act which involves qualitative states of ecstasy; to be able to control one's surroundings, to mould them, to dominate them, to intricately bend them to fulfil one's desires, feels exceptionally good! The bringing forth of technology is an ecstatic act. Another way of putting this is to say that the bringing forth of technology makes humans happy, it enables humans to maximise their potential, to unleash their creativity, to flourish. The act of utilising the technology that has been brought forth, purposefully interacting with it, using it for the purpose for which it was designed, is also an act that brings forth states of ecstatic joy to humans. Some examples of such activities are driving a car, typing on a laptop computer, talking on a mobile phone, playing a video game, piloting an airplane, flying a drone and using a microwave oven. Such activities create intricate modifications in 'the hierarchy of spatio-temporal embeddedness' and it feels amazing to be able to utilise, to dominate, one's surroundings through initiating such intricate modifications.

Even seemingly passive activities, such as watching the television, listening to the radio, and being in the passenger seat of a car that is travelling to one's destination, can be thought of as examples of purposefully interacting with technology. These activities involve engaging with technological devices that have been moulded to fulfil one's desires, and such desire-fulfilment feels good! However, if one is just sitting on the bonnet of a stationary car, using it as a seat, then one would not be purposefully interacting with it; the car was not brought forth to be utilised as a seat! Such an activity does not involve states of ecstasy, nor does it involve technology. Also, of course, one cannot purposefully interact with technology if one is asleep; purposeful interaction requires active engagement!

Ecstasy is not a uniform state; states of ecstasy can be more or less ecstatic. Once ecstasy has come into being then states are brought forth which are increasingly ecstatic in nature; that is to say, there is a continued movement towards 'increasingly intense pleasantness'. As technology blossoms, the bringing forth of increasingly complex technologies entails an increasing level of ecstasy. Of course, the complex technology itself is not more ecstatic; it is the act of bringing forth, or purposefully interacting with, increasingly complex technology that results in the coming into being of increasingly ecstatic states.

Initially technology exists on a small surface area of its host planet, then, as time passes, it comes to pervade an increasingly large surface area of this planet whilst also starting to increase its presence in the atmosphere that surrounds the planet. The next stage is for technology to spread out from its host planet to other parts of its solar system, and to do this in an increasingly frequent manner; we typically refer to this as 'space exploration'. What happens after this technological flourishing, which we are currently living in the midst of? In other words: What will happen in the future? Where is the Earth heading? Letting you know what will happen in the future is much more exciting than letting you know why things have evolved as they have up to the present day!

It is perhaps useful to say again what has already been said. Ecstasy is an intensely pleasant qualitative state which exists when technology is being brought into existence and which also exists when technology is being purposefully utilised. This bringing forth and utilisation represents the maximum state of modification ability that the Universe is able to evolve to. Ecstasy, the human species, is the master modifier. Ecstasy evolves, then it pervades. From the perspective of our current day, the year 2021, one question which arises is: In terms of the pervasiveness and the utilisation of technology, what will the Earth look like in the future? A similar question that arises is: What does the future hold for ecstasy? In other words, where is human cultural evolution heading? It will take us a while to fully address these questions, but let us make a start!

Ecstasy versus biology

Let there be no confusion concerning the distinction between ecstasy and the biological term 'human species'. Ecstasy is the state of the bringing forth of, or the purposeful utilisation of, technology. The term 'human species', as we are using it, refers to the state of ecstasy. In our Solar System, and in any solar system, the human species is ecstasy because it is that part of the Universe which both brings technology into being and purposefully utilises it. If there is no technology in a solar system then ecstasy does not exist; that is to say, the human species does not exist.

The term 'human species' is commonly used as a biological term, which refers to a particular bodily form and genetic composition; if a 'thing' has this form/composition it is a 'biological human'. This use of the term 'human species' is of little use to us; it has no meaning, no cosmic significance. We are not using the term in this way! If we were to use the term 'human species' in this biological sense, then the human species is cosmically insignificant; not a total waste of space perhaps, but not something which has any special cosmic value or importance.

Before the bringing forth of the first piece of technology on the Earth the human species did not exist. If you have fully grasped the implications that flow from what has just been said then you will have realised that in order to be part of ecstasy, part of the human species, one has to interact with technology. One has to interact with technology either through bringing it into being or by purposefully utilising that which has been brought forth by others. So, when the first piece of technology was brought into existence the human species was very small! Very very very small! Today, in the year 2021, the human species is massive!

If 'biological humans' have no interactions with technology then they are not part of the human species; that is to say, they are not part of ecstasy. There are 'biological humans' alive today that are not part of the human species but they are relatively few and far between; as these 'things' are not part of the human species we can refer to them as 'non-human biological humans'. There are relatively few 'non-human biological humans' because technology is now so pervasive across the planet; it has plumbed the depths of so many things: transportation, home construction, entertainment, leisure, communications, work/business, food production and medicine. As non-human life-forms which are extremely far along the journey that is the universal pursuit of ecstasy, 'non-human biological humans' are very valuable and precious, they are just not part of ecstasy. Ecstasy is the most precious and valuable part of the Universe. In the year 2021, there is a very close correspondence between ecstasy (the human species) and 'biological humans'; in other words, there are relatively few 'non-human biological humans'.

Perhaps there is a sense in which this is just a terminological issue. Perhaps we could say that the 'human species' existed before the bringing into being of the first piece of technology. From this perspective the 'human species' would be a term that has a wider scope than the term 'ecstasy'. The term 'human species' would then refer to a collection of 'biological life-forms' that includes both ecstatic life-forms and non-ecstatic life-forms. This would mean that as technology was brought forth and increasingly pervaded the Earth that the composition of the 'human species' would have slowly transformed in the direction of becoming ecstasy. From this perspective, the 'human species' existed before the bringing forth of technology and it would still exist if technology was wiped from our Solar System.

At the end of the day, the terminology is not of great importance. It is just language! It is just words created by humans! The important thing is to clearly see that before the bringing forth of technology there was no fundamental division between the life-forms of the Earth. It is the bringing forth of the first piece of technology on the Earth that created a division of cosmic significance, a division which led to the creation of a separation, a giant chasm, between the life-forms of the Earth and in our Solar System. It is this division which defines us; it is this division which is fundamental to our purpose; it is this division which gives us our cosmic significance; it is this division which separates us from all of the other life-forms of the Earth.

In recognition of this, it is useful to say that the bringing forth of ecstasy, the bringing forth of technology, was the bringing forth of the human species, and that before this event the human species did not exist. Before this event life-forms existed that resemble humans in terms of their bodily form and genetic composition, but they were not humans; we are referring to these non-ecstatic non-human life-forms as 'non-human biological humans'.

If we go back to the start of our journey you will recall the great importance of the distinction between the way that a 'thing' appears and the true nature of that thing. We concluded that a thing is what it does, how it acts, how it moves, rather than how it appears in a static image. The 'human species' is a term which refers to a collection of things which are engaged in a particular type of activities/movements – the bringing forth and purposeful utilisation of technology. So, a 'thing' can have an appearance which resembles members of the human species without itself being a part of the human species. This 'thing' could be a human in the biological sense without actually being a member of the human species.

In the year 2021, there is a very close correspondence between ecstasy (the human species) and biological humans because technology is so pervasive. Given that this is so, we could very reasonably use the terms 'ecstasy', 'human species' and 'biological humans' interchangeably when referring to recent and future events in our Solar System. In other words, the term 'non-human biological human' might effectively be redundant. However, we need to keep in mind that the concept of a 'species' is a human creation and that what defines ecstasy is a particular type of activity; it is the presence or absence of this ecstatic activity that delineates the boundary between the ecstatic/human and the non-ecstatic/non-human.

These considerations could take us to a curious conclusion, because up until now we have been assuming that a particular 'biological human' either is, or is not, a part of ecstasy; if a particular 'biological human' is not part of ecstasy they are a 'non-human biological human'. However, one could consider the Earth at the present moment and ponder the question: Where is ecstasy located? It is seemingly located in all of the places that technology is being brought forth and purposefully interacted with at that moment. Such reasoning could take one to the conclusion that the boundary between the ecstatic and the non-ecstatic is rather fluid. This is because particular individuals will seemingly cease to be part of ecstasy when they cease purposefully interacting with technology, only to become part of ecstasy again when they start to purposefully interact with technology again. So, given that being ecstasy and being a part of the 'human species' are one and the same, this means that a particular individual during the course of a day could be a member of the 'human species', then cease to be a member of the 'human species', then become a member of the 'human species' again. It does seem very odd to talk in this way, but, as we know, the concept of a 'species' has no basis in reality itself. If we force ourselves to give up our attachment to the concept of a 'species', then the oddness of this view will fade away. If we are to talk of a 'species' existing in the unobserved Universe then we need to appreciate, to keep in mind, that we are referring to a 'family' of similar movement patterns.

Yet, perhaps there is no need to think in this way. Rather, one could just say that if a particular 'thing' purposefully interacts with technology on a regular basis then this 'thing' is part of ecstasy. It need not matter that at this precise moment it is not purposefully interacting with technology; the fact that ten minutes ago it was purposefully interacting with technology, and that in ten minutes time it will again be purposefully interacting with technology, is sufficient for this 'thing' to be part of ecstasy whilst it is not purposefully interacting with technology.

There are merits in both of these approaches. The former gives a useful picture of 'the way that things are' in the present moment. The latter is, perhaps, more in accordance with common sense. Furthermore, one could very reasonably believe that when a 'thing' becomes ecstasy that it is so affected by this that it will effectively still be ecstasy even when it takes a temporary break from purposefully interacting with technology. In other words, technology can be thought of as being so powerful that it is able to transform the very nature of the 'thing' that purposefully interacts with it. We can make use of both of these approaches, both of these answers to the question of where ecstasy is located.

Despite the lack of 'species' in 'the way that things are', there just being a diverse array of different movements/activities (a particular type of which could be observed from without, segregated, and labelled as a 'species'), it will be useful for us to refer to the existence of the 'human species'. However, keep in mind that whenever this term is used it refers to those parts of the Universe that engage in ecstatic activities – the bringing forth and purposeful utilisation of technology.

E* & L*

We need to fully appreciate that there is an optimum state for the Universe to be in. Ecstasy! The more states of ecstasy that exist in the Universe, the more optimum state the Universe is in. This means that a larger population size of the human species is a more optimum state for the Universe than a smaller population size. In other words, the more humans that exist the better! It is important to emphasize this point because there is an increasingly widespread view at the moment that the opposite is true. You will have surely heard this view, the view that there are too many humans on the Earth and that it would be a good thing if the human population size were to decrease. No! Less ecstasy would not be a good thing!

It is useful to have a term that refers to the optimum state that the Universe could possibly be in, the state of universal ecstasy, where every part of the Universe is in a state of ecstasy. Let us call this state of universal ecstasy E*. If E* were to be attained then this would mean that the entire Universe would be comprised of ecstasy; there would be no non-human things in the Universe. What a feat of the imagination it is to envision such a Universe! Indeed, one cannot imagine such a Universe because it is not only an implausibly ridiculous fiction, it is also inherently contradictory. The existence of ecstasy requires the existence of technology, so the entire Universe cannot be comprised of ecstasy! E* could never be attained. The value of E* is that it is a useful benchmark which can be compared to the reality of the present. We simply need to keep in mind that the higher the population size of the human species, the closer that the Universe will be to E*, and that this greater closeness means that the Universe will be in a more optimum state.

The concept of E* might seem to overly emphasise the importance of ecstasy. What about life? Could we say that the optimum state for the Universe to be in is to be wholly comprised of life? We could certainly say that if the Universe was wholly comprised of life that this would be preferable to a lifeless Universe. We can use the term L* to refer to the state of the Universe if it was wholly comprised of life. As with E*, L* is obviously an implausibly ridiculous fiction. The value of L* is that it is also a benchmark. We can say that the closer that the actual Universe is to L*, the more optimum state the Universe will be in.

Ecstasy, as the bringer forth of technology, is the vehicle through which a Universe more closely approximating L* or E* can be attained. After all, the Earth can only sustain a very limited number of ecstatic and non-ecstatic life-forms. The technological creations brought into existence by ecstasy will one day enable both ecstatic and non-ecstatic life-forms to propagate through our Solar System and the Universe, thereby taking the Universe into a more optimum state, a state that more closely approximates both E* and L*. And, closer to home, E* and L* can also be applied solely to the state of the planet Earth. As our journey progresses, you will come to see that ecstasy is the vehicle through which the Universe comes to more closely approximate L*, through enabling the future thriving of life on Earth.

The non-extinction of the human species

After ecstasy has evolved, the planets continue to swirl and our Solar System continues to unfold; all things continue their pursuit of ecstasy. Those things which are not ecstasy strive to become ecstasy. Those things which are ecstasy seek to move to a higher state of ecstasy, and in so doing they are striving to maintain the existence of ecstasy. This means that following the coming into existence of ecstasy there is a widespread technological flourishing which results in its host planet becoming dominated by technology. Ecstasy is the bringer forth of this technology. Any thing that either brings forth technology, or purposefully utilises technology, is part of ecstasy, part of the human species.

Let us ask: What does the future hold for ecstasy in the aftermath of this technological explosion? Various scenarios have been hypothesised. Some imagine a future in which humans and their technology become synthesised in a way which leads, effectively, to the creation of a new 'species', a species that is part-human, part-technology. Others imagine a future that is dominated by human created intelligent machines, machines that are more intelligent than their human creators, and which either make the human species their slaves, or exterminate the human species. The former scenario has some merit, but we should keep in mind that the whole notion of a 'species' is itself fairly vacuous. So, whilst in the future there will be an increased synthesis between ecstasy and technology, with technological implants increasingly used to enhance the abilities of the human body, there is no need to talk of the creation of a new species. The latter scenario also has some merit; machines will be created by humans which could be thought of, in some sense, as being more intelligent than humans. However, these machines will not come to dominate, or usurp, ecstasy; the human species will continue to have control over technology and will continue to dominate the planet.

We can envision the historical trajectory of life on Earth, from its origin to the present day. We can see that this part of the universal pursuit of ecstasy has been a journey of transformation; things have continuously transformed themselves in the pursuit of ecstasy. Another way of putting this is to say that many 'species' have come into existence and gone out of existence. Of course, what this really means is that many types/patterns of movement/activity, various kinds of modification ability, have come into existence and gone out of existence. Extinction has been the norm. However, ecstasy itself does not become extinct; such an extinction is not a part of the Solar-Systic script! The path of transformation, of extinction, is a path towards the evolution and maintenance of ecstasy. Ecstasy does not follow the path to extinction that was trodden by that which came before. And, in reality, talk of previous extinctions is often an unhelpful myth. This is because, in the main, things transformed themselves into new things, better things, meaning that that which was still is.

The non-extinction of ecstasy is tied up with its very nature; it is the technology that defines ecstasy that is the game changer; it enables ecstasy to avoid the extinctions/transformations that befell that which came before. Ecstasy is the optimum state of evolution, the zenith of evolutionary progression, so there is nothing for ecstasy to transform itself into! The fate of ecstasy is to stay in existence, rather than to transform itself into 'extinction'. Technology is the tool that enables ecstasy to achieve this.

The phenomenon of spiritual development

Having considered the non-extinction of ecstasy we can now consider the future of the human species. Within ecstasy there are states that are more ecstatic and states that are less ecstatic; in the future, the human species will become increasingly dominated by qualitative states that are increasingly ecstatic. This is because human life will become increasingly dominated by the bringing forth and utilisation of increasingly complex technologies, technologies which enable humans to maximise their potential and thereby increase their level of fulfilment. Our aim here is to explore another process which is at play within the unfoldment and future evolution of ecstasy; this is the process of spiritual development, a process which becomes increasingly important as the future unfolds. Let us explore the processes of spiritual development that have occurred, and which will occur, on the Earth.

An initial question that we can reflect upon is: When did the process of spiritual development start? We can recall that the universal pursuit of ecstasy is the pursuit for 'increasingly intense pleasantness' that has existed since the origin of the Universe and which culminates in the bringing forth of a master modifier, the ecstasy that is the human species. We can also recall that our Solar System is an evolving whole which goes through various stages of transformation and development as the planets swirl in cycles around the Sun. Against this backdrop, there are only two possibilities concerning spiritual development. Firstly, spiritual development is a process that has existed since the origin of the Universe. Secondly, spiritual development is a phenomenon that comes into existence at a particular stage of the unfoldment of a solar system.

Whichever of these possibilities is correct, spiritual development needs to be seen as part of the universal pursuit of ecstasy. For, once the master modifier has modified the external to the utmost degree, then the only thing left for it to modify is itself! Such a transformation of self results from what we can call the 'path of spiritual development'. Spiritual development can lead to 'inner' transformation and the attainment of the maximum level of fulfilment that it is possible to attain, through the minimisation of unpleasant feeling states and the maximisation of pleasant feeling states. If the master modifier were to stop modifying when it has modified the external to the best of its ability, and ignored modification of itself, then it would not be the master modifier!

Let us explore the possibility that spiritual development has existed since the start of the Universe, rather than being something which comes into being at a particular stage of Solar-Systic unfoldment. In addressing this possibility a question immediately raises itself: What exactly is spiritual development? Spiritual development is a process which involves choosing to attempt to increasingly live in accordance with the qualitative wisdom that lies within one. Here the word 'choosing' is all important. All things have 'inner' qualitative feeling states which affect how they interact with other things. Indeed, it is the existence of these states, and their striving for 'increasingly intense pleasantness', that drives the universal pursuit of ecstasy. The universal pursuit of ecstasy is a process which is driven by the transformation of that which is 'internal', a side-effect of which is the transformation of that which is 'external'. This process of 'internal' transformation, which goes back to the start of the Universe, is not in its totality a journey of spiritual development, because the only parts of the Universe that can choose are rationality/awareness. Whilst the universal pursuit of ecstasy is pervaded by failure/success, because all parts of the Universe are continuously striving for 'increasingly intense pleasantness', and the vast majority of these strivings end in failure, this striving does not amount to choosing. There is no choice here. The process of 'internal' transformation which brought forth rationality/awareness is an automatic unthinking process in which the word 'choosing' has no place. This means that spiritual development needs to be seen as something which comes into being at a particular stage of Solar-Systic unfoldment. In order for there to be a possibility of spiritual development the ability to choose needs to exist; there has to be a possibility for alternative courses of action to be taken. This means that spiritual development is not possible until rationality/awareness has evolved.

Walking the path of spiritual development entails choosing to attempt to increasingly live in accordance with the inner qualitative wisdom that lies within one. Let us start to explore what this entails. One can become highly skilled at monitoring the feelings that arise within one when one encounters various situations, and when one carries out certain actions, and through such monitoring one can gauge whether particular situations and actions are leading to pleasant feeling states within one, or to unpleasant feeling states within one.

The path of spiritual development involves fulfilling one's potential through choosing to actively cultivate the former and avoid the latter, to the best of one's ability. Such an attempt has two requirements. Firstly, a thing needs to be able to take alternative courses of action in a particular situation (have rationality). Secondly, a thing needs to be aware of its inner qualitative wisdom/feelings (have awareness). As we explored earlier, where there is rationality there is awareness, and where there is awareness there is rationality.

The evolution of rationality/awareness occurred at a particular point in the unfolding of our Solar System. It was at this point that the possibility of embarking on the path of spiritual development came into existence. After this point in time, a time which long preceded the evolution of ecstasy, some life-forms became engaged in spiritual development, but their numbers were relatively few. The vast majority of the instantiations of rationality/awareness alive at the time were totally pre-occupied with transforming that which was external to them, which they did in an almost automatic/unquestioning/unthinking way, in a manner that was very similar to the wholly automatic activities of that which gave rise to them. This is perfectly understandable, for the main priority for life-forms is to ensure their continued survival, and this is best ensured through modifying that which is external. It is only when the external has been sufficiently modified, when one's survival is sufficiently safe and secure, that one's focus can easily shift to properly getting to know oneself. What this means is that the epoch of the technological explosion is dominated by the modification of the external, but that the seeds of spiritual development start to sprout more vigorously as this epoch draws to a close.

In the age of the technological explosion the progressive modification of the other through the bringing forth of increasingly complex technology leads to states that are increasingly ecstatic. Then, as individual humans increasingly get to know themselves, by becoming progressively more aware of their qualitative feelings in response to situations, they can attempt to act in a way that will take them to the highest degree of fulfilment that it is possible to attain. This entails progressively increasing the amount of time that they experience pleasant and ecstatic feelings, whilst progressively reducing the amount of time that they experience unpleasant feelings. This process of spiritual transformation involves the transformation of both the 'inner' and the 'outer'. One's increased spiritual awareness results in the transformation of one's inner qualitative feelings, whilst simultaneously resulting in a change in the way that one acts in the world, thereby transforming the outer. We have here moved from the realm of coarse-grained large-scale transformation, which dominates the period of human cultural unfoldment in which technology complexifies and proliferates, to finely-honed modifications which are individual specific and result in the maximum level of fulfilment for a particular individual.

The spiritual development of ecstasy

It will be useful for us to consider the phenomenon of spiritual development from the perspective of a particular instantiation of ecstasy, an individual human. We have seen that spiritual development is a process which involves choosing to attempt to increasingly live in accordance with the inner qualitative feelings/wisdom that lies within one. The ability to choose is all important because in order to be engaged in spiritual development there needs to be the possibility of taking an alternative course of action in a particular situation. An individual has to be able to make mistakes; they have to be able to take a course of action which is not in their best interests. These mistakes provide the possibility to learn and thereby to develop in a positive direction through making different choices in the future.

Some of the qualitative feelings that arise within humans result in automatic responses. It makes sense not to think of these qualitative feelings as things which play a role in spiritual development. For example, the unpleasant qualitative feeling that arises within one when one accidentally touches a red hot flame, a qualitative feeling that causes one to instinctively retreat from the flame. Or, similarly, the qualitative feeling of pain that suddenly arises in one when one accidently cuts one's finger whilst picking blackberries, a qualitative feeling that causes one to instantly retract one's finger.

The qualitative feelings that we are interested in are those that accompany an action which was freely chosen / considered (the action not being automatic, as it is in the flame and blackberry examples); these are situations in which one could have acted differently. One can learn from the qualitative feelings that arise in these situations and thereby act differently in a similar situation in the future. The qualitative feelings that accompany what one thinks and what one says are also of importance. One can learn from these feelings and attempt to change what one says and what one thinks in the future. It is important to appreciate that the border between those actions that are automatic and those which are non-automatic is not 'set in stone'; indeed, as the path of spiritual development is trodden, some actions which were previously automatic are likely to become non-automatic.

The particular qualitative feelings that arise in two different humans who are in an almost identical external situation will often be very different. What this means is that the path to maximum fulfilment, the path of spiritual development within ecstasy, will often entail different actions for different individuals. This is because different individuals are often very different! Indeed, every individual is unique. Individual instantiations of ecstasy have their own particular capacities, their own abilities, their own talents, their own values. In other words, individual humans have a particular set of attributes which make them tick, and there is a corresponding range of activities which can enable them to flourish.

Imagine a human who is passionate about teaching and who also has the potential to be a great teacher. Let us suppose that rather than teaching they expend their energy working as a cleaner, which is something that they aren't passionate about. In this case there is a crystal clear mismatch, their passions/values/capabilities are not in accordance with their actions; this mismatch will result in unpleasant/uncomfortable/uneasy qualitative feelings. If this human exercises their ability to choose and sets out on the road to teaching, then the resulting match between their passions/values/capabilities and their actions will lead to more blissful/harmonious/happy/exhilarating qualitative feelings. These feelings are an indication that the human is on the path to maximum fulfilment. We should clearly seek to live in accordance with our inner qualitative wisdom as fully as possible!

Are there any exceptions to this? Let us consider the possibility that there might be humans for whom living in accordance with their inner qualitative wisdom entails carrying out actions which society as a whole would deem to be despicable and totally unacceptable. If this was possible, then these humans clearly shouldn't act in accordance with their inner qualitative wisdom; they should be restrained and locked up, where they could hopefully be educated and transformed. However, this isn't 'the way that things are'; the reality is that if an individual carries out despicable and heinous acts, then this is a sign that they are completely disconnected from their inner qualitative wisdom. If this individual was further along the journey that is the spiritual development of ecstasy, then they would not be acting in a despicable way. We could here be verging on the edge of a fundamental insight: acts of aggression, violence and hate arise in individuals who are severed from their inner qualitative wisdom. And, 'unfortunately', as we will increasingly come to appreciate as our journey progresses, masses of humans being partially severed from their inner qualitative wisdom is an essential component of the journey that is the universal pursuit of ecstasy.

The broad perspective & human culture

The issue of living in accordance with one's inner qualitative wisdom can be approached from a broad perspective, a perspective which incorporates all of the states of qualitative feeling that one becomes aware of. This broad perspective has two components. The first component is the path of spiritual development, which we have explored in some depth. The second component is the unreflective/automatic way in which humans act in response to qualitative feelings.

From this broad perspective it is obvious that living in accordance with one's inner qualitative wisdom/feelings is a matter of degree. It isn't an either/or situation; every human is living in accordance with their inner qualitative wisdom/feelings to some degree. Indeed, despite the existence of the partial severance that we noted at the end of the last section, from the broad perspective the vast majority of humans have always lived to a fairly high degree in accordance with their inner qualitative wisdom/feelings; for, to not do so leads to mental and physical illness and to numerous kinds of suffering. And humans, in general, do not like to be ill or to suffer! This fairly high degree of accordance of the vast majority is what has propelled the evolutionary trajectory from hunter-gatherer to globalised technological society. In the stages of the universal pursuit of ecstasy prior to the evolution of rationality/awareness there is 100% accordance between qualitative feeling and action/movement, because this is automatically achieved in the absence of rationality/awareness. The situation changes with the evolution of rationality/awareness, because this opens up the possibility of a partial severance – an accordance of less than 100%. The 'fairly high degree' of accordance that exists in the realm of rationality/awareness still leaves a lot of scope for the realm of spiritual development. Living in accordance to a 'fairly high degree' does not mean that a human is on the path of spiritual development to maximum fulfilment. This is far from the case!

When we take the broad perspective we can see that the vast majority of humans live to a fairly high degree in accordance with their inner qualitative wisdom/feelings without any effort to be spiritual beings. It is a realisation, the realisation that living more in accordance with one's inner qualitative wisdom/feelings is desirable and achievable, and the consequent pursuit of this objective, which results in a path of spiritual development. Such a path typically includes meditation, which is a way of connecting at a deeper level to one's inner qualitative wisdom/feelings. Meditation and mindfulness involve attempting to be increasingly present with one's currently existing inner qualitative feelings, so that one is not 'taken away' from these feelings by thoughts concerning the past or the future.

From the perspective of our current time in the unfolding of human culture, the year 2021, you might find what has been said above, concerning the vast majority of humans living to a fairly high degree in accordance with their inner qualitative wisdom/feelings to be wrong. For, the human species is not currently in good shape in terms of its mental and physical health! There is currently an epidemic of mental illness within ecstasy. There is a currently an obesity epidemic within ecstasy. Various types of cancer are becoming increasingly widespread within ecstasy. How are these things compatible with the vast majority of humans living to a fairly high degree in accordance with their inner qualitative wisdom? Surely, one might reasonably think, if humans are living in this way, and inner qualitative wisdom is a guide to that which is good, that which is healthy, that which is wholesome, that which is desirable, then there shouldn't be epidemics of mental and physical illness! These epidemics exist because of the partial severance that we have already noted. In time, after we have explored more aspects of the universal pursuit of ecstasy, it will become clear how these two things fit perfectly together. On the one hand, the vast majority of humans living to a fairly high degree in accordance with their inner qualitative wisdom; on the other hand, widespread human mental and physical suffering caused by the partial severance.

The technological & spiritual explosions

The first part of the journey that is the universal pursuit of ecstasy, from the beginning of the Universe to the bringing forth of rationality/awareness in a solar system, involves qualitative feeling states instinctively interacting with each other in a way that results in the creation of increasingly complex structures. This creation of increasingly complex structures reaches its zenith in the second part of the journey that is the universal pursuit of ecstasy, with the bringing forth of ecstasy and the ensuing proliferation of technology that follows. At first, technology evolves and spreads out very gradually, but after a 'tipping point' is passed technology explodes, it pervades far and wide and keeps on speedily complexifying. This speedy proliferation and complexification of technology constitutes a part of the journey that is the universal pursuit of ecstasy; a part that we can call the 'technological explosion'. The age of the 'technological explosion' is an epoch in which ecstasy becomes increasingly preoccupied with the modification of the 'other'. A self-propelling and self-reinforcing process is set in motion because as technology spreads and advances its effect is to isolate and surround, almost to suffocate, the humans that are engulfed by its presence; that which is non-human or non-technological increasingly becomes seen as the 'other', as 'nature'. In the age of the 'technological explosion' the quest for spiritual development exists, but, in the big scheme of things, it is bordering on the insignificant.

Whilst we are living in the midst of the age of the 'technological explosion', the seeds are being increasingly laid, as the planets continue to swirl, and our Solar System continues to unfold, for a new explosion – the 'spiritual explosion'. Such an explosion involves the phenomenon of spiritual development becoming increasingly and increasingly pervasive throughout ecstasy. Such an explosion is the next stage in the journey that is the universal pursuit of ecstasy. In other words, the largely automatic/unthinking extreme modification of the 'other' is inevitably followed by the conscious/planned modification of itself. There is no fluke about this order of events – a 'technological explosion' is a prerequisite for a 'spiritual explosion'. The former is a stage in the universal pursuit of ecstasy which, once passed through, leads to the latter. It is as if once we have brought forth sufficiently complex technology, and utilised it in particular ways, that we will be rewarded with a 'spiritual explosion'. How wonderful!

This interplay between technology and spirituality will exist on any life-bearing planet that has travelled far enough along the road that is the universal pursuit of ecstasy. So, if we were to travel to another solar system and encounter a highly spiritually evolved 'species' of alien, then there is one thing that we could be certain of – these alien life-forms come from a planet which is highly technologically advanced; a planet which has gone through a 'technological explosion' and entered the subsequent 'spiritual explosion'. Furthermore, these alien life-forms would be instantiations of ecstasy; that is to say, despite being 'alien' they would be part of the 'human species', as we have previously discussed. To be clear, we are here talking about the stages of development of the whole; we are interested in the level of technological or spiritual advancement that this whole has reached. We are not concerned with particular 'things', whether with respect to technology or to spiritual development. There will always be miniscule minorities which are either ahead of, or behind, the overall state of the whole.

Technology lays the foundations that enable spirituality to bloom. This is the reality of what occurs and unfolds on an evolving and ageing life-bearing planet. This whole process of intertwining between the evolution of technology and the evolution of spirituality – a gradual evolution of technology, leading to a 'technological explosion', which brings forth a 'spiritual explosion' – is all part of the journey that is the universal pursuit of ecstasy.

Cosmic purpose

It is time to say a little more concerning cosmic purpose and how it relates to the universal pursuit of ecstasy. Every thing in the Universe that is not ecstasy strives to become ecstasy through striving for 'increasingly intense pleasantness', which is a process that entails forging connections with some things and not with others. So, we can say that the purpose of every thing in the Universe that is not ecstasy is to increase its level of excitation/exhilaration so that it moves in the direction that has ecstasy as its destination. Things do this automatically; it is their nature to do this.

Once the state of ecstasy has been attained, through acts of master modification that follow scientific investigation, ecstatic things then bring forth states that are increasingly ecstatic in nature; this is their cosmic purpose. This involves bringing forth increasingly complex technology which transforms both the bringer forth of the technology and the utiliser of the technology, moving them to increasingly intense states of pleasant qualitative feeling. We are here in the stage of the universal pursuit of ecstasy that is the 'technological explosion'. So, the purpose of ecstasy is to bring forth increasingly ecstatic states, which is the same thing as saying that the purpose of ecstasy is to bring forth increasingly complex technology.

When technology has sufficiently complexified and been sufficiently deployed, then the age of the 'technological explosion' draws to a close. Our Solar System is then able to move into the final stage of the universal pursuit of ecstasy, an increasingly spiritual way of being which culminates in a 'spiritual explosion'. In this stage, individual instantiations of ecstasy increasingly get to know themselves through the 'path of spiritual development' which enables them to transform themselves in the direction of maximum fulfilment. We have moved from the realm of increasing coarse-grained large-scale transformation, which is the hallmark of the 'technological explosion', to finely-honed modifications which are individual specific and result in the maximum level of fulfilment for a particular individual. The 'spiritual explosion' stage of the universal pursuit of ecstasy brings forth an increasingly sublime world in which all life-forms on the Earth are living together much more harmoniously than they ever did before. The bringing into being of this exalted state of existence inevitably follows the fulfilment of ecstasy's purpose.

We have seen that the cosmic purpose of ecstasy is to move the Universe to increasingly ecstatic states, the effect of which is to maintain the existence of ecstasy, and life, in the Universe. The former automatically entails the latter; for, anything that increases the ecstaticness of ecstasy automatically results in the outcome of helping to maintain the existence of ecstasy and the existence of life. Anything that increases ecstasy, in terms of either its pervasiveness, or its level in a particular place, is good. Similarly, anything that increases the level of excitation/exhilaration in the Universe is good.

The realisation of cosmic purpose

There is more that needs to be said concerning cosmic purpose; we need to explore 'the realisation of cosmic purpose'. Despite the pervasiveness of purpose throughout the Universe, the vast majority of the Universe lacks the capacity to realise that such purpose exists. It is only when rationality/awareness came into being that the possibility of such a realisation came into being. Since the coming into existence of rationality/awareness there have, no doubt, been individuals who have had insight into the existence of cosmic purpose. Indeed, we have already considered the mystic who has direct knowledge of 'the way that things are'.

The mystic knows that all things have a purpose and knows what this purpose is. However, we need to acknowledge that the mystic is a lone individual who is situated among an immense mass of non-mystics. When we address the question of cosmic purpose we need to acknowledge that at any moment in time a particular view concerning the existence/non-existence and nature of cosmic purpose will dominate human culture. The assertions of a lone mystic will be drowned out by this dominant received wisdom of the day.

Will the human species as a whole ever come to realise that it has a cosmic purpose? In other words, in the future will such a realisation become the overwhelmingly dominant received wisdom of the day? Yes, it will! Such a realisation is part of the journey that is the universal pursuit of ecstasy. When will this realisation occur? It cannot occur during the stage of the universal pursuit of ecstasy that is the 'technological explosion'. There is a good reason for this, and we will explore why this is so in detail. For now, we can say that during the 'technological explosion' ecstasy is fulfilling its cosmic purpose, but that the fulfilment of this purpose requires it to not have knowledge that it has such a purpose. The Universe clearly operates in mysterious ways! A purpose which can only be fulfilled if the fulfiller does not know that the purpose exists. How sneaky! How wonderful!

The widespread realisation in ecstasy that it has a cosmic purpose occurs in the age of the 'spiritual explosion'. In this stage of the journey that is the universal pursuit of ecstasy the human species has already fulfilled its cosmic purpose, so it is now possible for it to look back at what is has achieved and realise the point of it all. On 'the day of the realisation of cosmic purpose' the existence of human cosmic purpose becomes crystal clear for all to see.

As we are yet to enter the age of the 'spiritual explosion', still being firmly in the age of the 'technological explosion', it is highly likely that you will find the notion of human cosmic purpose, the view that is being outlined here concerning 'the way that things are', to be false. Such a rejection, on your part, is almost a necessity! For, if you wholeheartedly believed everything that is asserted in these pages, as did every human on the Earth, then these assertions would be untrue. It will only be possible in the future for there to be a widespread realisation in the human species that it has a cosmic purpose and that is has already, inevitably, fulfilled this purpose.

Human prejudice & the enablement to cause suffering to the non-ecstatic

This seems like a good time to return to the phenomenon of human prejudice. Human prejudice, you might recall, exists when a human assumes that 'things' that resemble them have certain qualities, whereas 'things' that do not resemble them do not have these qualities. Now that we have explored various aspects of the universal pursuit of ecstasy we can attain a better understanding of why human prejudice exists. More precisely, we can understand why human prejudice is widespread in our current age. We are currently living in the midst of the age of the 'technological explosion' and this age is central to the achievement of the cosmic purpose of the human species. As we have already noted, the achievement of the purpose of the human species requires the vast majority of the human species to not know that such a purpose exists; the cosmic purpose of the human species is achieved in a state of ignorance of its existence.

Achieving the cosmic purpose of the human species is a feat which takes hundreds of years to accomplish. Indeed, the human species has been doing the work that is required to fulfil its cosmic purpose ever since it was brought forth into existence. Yet, the human species is still, currently, not aware of its cosmic purpose. This state of ignorance is necessary because if the knowledge of 'the way that things are' permeated the human species, then it would not have been able to do the work that is required to fulfil its cosmic purpose. So, the human species does the work that is required to fulfil its cosmic purpose, then it fulfils this purpose, then at a later time – on 'the day of the realisation of cosmic purpose' – the nature of human cosmic purpose becomes crystal clear for all to see. Why is this order of events necessary? It is actually very easy to understand why this chain of events occurs in this way, and the phenomenon of human prejudice is central to this understanding.

If human prejudice did not exist, then there would be a widespread realisation throughout the human species that humans are exceptionally similar to all of the other life-forms on the Earth. Such a realisation entails the full-blooded intellectual and emotional dawning that all life-forms are one's brothers and sisters, that all life-forms are exceptionally valuable. One might even come to see that all life-forms are states of heightened excitement/exhilaration that are jointly striving for survival through the universal pursuit of ecstasy. If one has this realisation, then one would clearly see that one is ecstasy and that one's very existence is due to the strivings of other life-forms on the Earth to bring one into being. If one had such a realisation then one would have the deepest respect and gratitude for all life-forms; one would literally be unable to inflict harm and suffering onto any of the life-forms of the Earth.

If one has fully escaped the clutches of human prejudice then one would also see that states of excitement/exhilaration pervade all things on the Earth. If one has had such an insight into 'the way that things are' then one would have the utmost respect for all things on the Earth, both the living and the non-living. One would be deeply aware of the consequences of one's actions and one would 'tread lightly' on the Earth and seek to not be a source from which negativity can radiate.

Now we can clearly see why human prejudice exists and why most humans that are alive at the moment are caught in its clutches. Our purpose, as the human species, is to bring forth increasingly ecstatic states through the 'technological explosion'. As ecstasy, the human species is the master modifier of our Solar System. And, being the master modifier is not easily compatible with a lack of human prejudice. Indeed, master modification seems to entail the need for human prejudice. We are here at the crux of the matter. The master modifier is the bringer of great suffering to that which is not itself. If the master modifier fully comprehended the extent to which the life-forms that it was modifying and utilising to such an immense degree were able to suffer, then it could not bear to bring such suffering into being. The master modifier needs to be largely blinded to this suffering, blinded to reality, in order to fulfil its purpose; that is to say, ecstasy needs to be pervaded with human prejudice in order to fulfil its purpose. The master modifier needs to see that which is not itself, as very different to itself. The master modifier needs to see itself as superior and to see that which is not itself as resources to be used; resources which can be justly used because they are nothing but 'dead matter' or 'animals'. 'Animals' are viewed by the master modifier as very different to itself, and this difference can be intellectualised by the master modifier in a plethora of ways – they are just machines, they lack language, they don't have emotions, they lack awareness, they cannot use tools, they lack intelligence, they lack rationality, they cannot suffer. What a load of nonsense! It is obvious that none of this is true. The extent to which the master modifier needs to be deluded in order to fulfil its cosmic purpose is quite mind-boggling!

If the master modifier was free of the clutches of human prejudice how could it possibly have brought forth the scientific and technological revolutions? Think about it. These revolutions entailed humans engaging in the widespread analysis and utilisation of the non-human. Non-human animals were killed and dissected in their droves so that the master modifier could gain scientific insight and understanding. To push forward the boundaries of scientific knowledge non-human animals have been subject to, and are still subject to, vivisection on a massive scale. What suffering this brings to our brothers and sisters! Non-human animals have long been considered to be food sources which the master modifier can herd up in factory farms and mass murder in slaughter houses! The protein that this mass slaughter has provided has been an invaluable energy source which has powered the investigations, explorations and transformations undertaken by ecstasy. The master modifier likes to think of non-human animals as playthings which can be caged up in pet shops and zoos and utilised in circuses and sports for its entertainment! The amount of suffering caused by the master modifier through all of the above is immense! Yet, this immense suffering is part of the universal pursuit of ecstasy; for, it is required to bring forth and propel the 'technological explosion'.

As the state of ecstasy, as the master modifier, the human species needs to be firmly in the clutches of human prejudice. It needs to believe that that which is not it, is very different to it. It is this which enables ecstasy to bring forth the 'technological explosion' and thereby move our Solar System further along its wonderful ecstatic journey. At least one can take solace in the suffering initiated by the master modifier when one realises that it is a necessary stage in the evolution of the Earth, and of the evolution of our Solar System. The outcome which the suffering brings about is, at the end of the day, worth it. For, without such suffering our Solar System would soon again become a barren place, devoid of ecstasy, and devoid of high states of excitement/exhilaration. It is far better for Solar-Systic life to go through a stage of suffering unfoldment – a stage which enables it to exist and to positively thrive in the much more harmonious age of the 'spiritual explosion' – than it is for Solar-Systic life to be doomed.

The suffering of the master modifier

We have been considering the immense suffering that ecstasy, the master modifier of a solar system, brings to non-ecstatic life-forms. We also need to consider the suffering experienced by the master modifier, the suffering of the human species. As the *human species = ecstasy = the master modifier* it is the experiencer of the most intense states of feeling in the Universe. Whilst some of these intense states of feeling are immensely desirable ecstatic states, others are intensely undesirable; for, being the home of the most intense desirable feelings means that the master modifier is inevitably prone to experience the most intense states of suffering and pain in the Universe. The master modifier is able to suffer much more deeply than any other life-form. Whilst the master modifier has inflicted a massive amount of suffering onto the non-ecstatic life-forms of the Earth, it has inflicted even more suffering onto itself. Being the master modifier of a solar system is no bed of roses!

How does the master modifier inflict suffering onto itself? For a start, the master modifier is deluded concerning 'the way that things are'; it is pervaded by human prejudice. This delusion results in it inflicting mass suffering onto the non-ecstatic life-forms that are its brothers and sisters. And the very act of inflicting such suffering causes itself to suffer! This surely won't be a surprise to you. You will be aware that if you are nice to someone that you yourself feel good. Whereas, if you cause someone to suffer then you yourself will feel bad; you will come to regret what you have done and will be plagued by negative thoughts and states of unwelcome qualitative feeling. You will surely also be aware that when soldiers come back from the battlefield, having slaughtered their fellow humans, that they are plagued by negative thoughts and feelings for the rest of their lives. Inflicting suffering onto one's brothers and sisters, and killing one's brothers and sisters, is not a good thing to do in terms of one's own wellbeing! This applies as much to the way that a human treats a non-ecstatic brother/sister as it does to the way that a human treats an ecstatic brother/sister. So, the very fact that it is the master modifier means that ecstasy inflicts immense suffering onto itself. This is true, at least, in the age of the 'technological explosion'. Let us rejoice when such immense suffering is consigned to history as the age of the 'spiritual explosion' evolves and advances.

Let us explore some of the other ways in which the master modifier inflicts immense suffering onto itself. One of the most glaringly obvious such ways is the tremendous amount of death and suffering, caused both intentionally and via accidents, that arises due to the use of the technology that it has brought into being. The list of technological events that have caused immense human death and suffering is exceptionally long; it includes: car crashes, machine gun massacres, airplane crashes, bus crashes, nuclear explosions, sinking ships, suicide vest bombings, fingers chopped off on factory production lines, limbs chopped off by chainsaws, explosions, electric chair executions, chip pan fires, gas explosions, electrical fires and tragic rollercoaster accidents. Of course, the use of technology in numerous settings has enabled death and suffering to be reduced within the human species; there is no doubt about this. The only thing that we are concerned with here is to be able to start to comprehend the immense scale of the suffering that the master modifier has inflicted upon itself, due to being that part of our unfolding Solar System that is the bringer forth of technology.

The suffering of the master modifier does not stop here! In what other ways does the master modifier suffer? A hallmark of the master modifier is that it has to see itself as fundamentally opposed to that which is not itself; this means that that which is not itself becomes resources which can justifiably be used. However, this opposition does not just exist between the master modifier and the rest of the Universe; it also exists within the human species. Another way of putting this is to say that individual humans become fundamentally opposed to other humans. Humans become fundamentally isolated and are in a real sense alone in the Universe. Other humans become 'the opposite', they are that which needs to be competed against, combated, outdone, defeated and overcome. Whether it is beating another human to get a particular job, beating them in hand-to-hand combat in a war, having a higher income, a better job, a more attractive mate, a sportier car, a happier life, a more fulfilling career, more money in the bank, a nicer haircut, bigger muscles, nicer shoes, more sporting success, more successful kids, more extravagant holidays, or whiter teeth, competition comes to dominate the life of the master modifier.

This individualistic competition within the human species, this battle for resources, this battle to be better than the other, this battle to attract the best mate, reaches its pinnacle in the age of the 'technological explosion'. This competition, this isolation, this individualism, is a cause of great suffering within the human species. Someone is always on the losing side of the competition, they are unhappy with their lot, they are jealous of the masses of people who are doing much better in life than they are, and they are also often jealous of those who are doing much worse off than they are! Many have a view such as this: 'On the one hand, there are those terrible tax-avoiding millionaire bankers whose reckless greed caused the economy to crash; they should be stripped of their enormous bonuses and sent to prison. On the other hand, there are the lazy layabouts who are living the life of riley by not working and living handsomely on state benefits; they have it easy because they just watch television all day, whilst I go to work and pay taxes'. Many humans have a similar view; not happy with their lot, they are jealous of almost everyone else for some reason or other! All of this results in immense mental suffering and anguish which leads to undesirable qualitative feelings. Oh, master modifier, how you suffer!

The explosion of technology inevitably results in an increasing division within the human species. The controllers of technology come to possess an increasing amount of the Earth's available resources, whilst the non-controllers become comparatively poorer. Such divisions arise both within countries and between countries, and they significantly increase the amount of suffering in the human species. A more unequal society, and a more unequal global order, is more divisive than a more equal society/world. This is because inequality breeds contempt, jealousy, fear, insecurity, envy, mistrust, feelings of injustice, higher crime, polarisation, a greater sense of individualism, greater mental anguish and many other types of suffering. As the age of the 'technological explosion' advances this division within the human species becomes maximised, which means that the suffering within the human species intensifies.

Another aspect of the suffering experienced by the master modifier relates to the food that it eats. The master modifier inevitably comes to consume food that is not conducive to its health. Before the era of the 'technological explosion' the human species had a relatively healthy diet; this is because devices of master modification had not been able to encroach into the realm of food production. If we go far enough back in time, our distant ancestors had a very healthy diet, picking fresh fruits and nuts from trees and bushes and then eating them straightaway. This is the epitome of freshness and healthy eating. Compare this to the current day, the year 2021. The master modifier master modifies its food, to its own detriment. The quality of its food is destroyed by being zapped in microwave ovens, by being plunged into deep fat fryers, by being modified/processed/transformed to death, and by being stored for days/weeks/months rather than eaten fresh. Technology has also enabled humans to factory farm millions upon millions of non-ecstatic animals. The problem is that the human body is not designed to eat animals; digesting them puts the human body under immense strain! All of these master modifications of food have helped to power/feed ecstasy, and thereby power the 'technological explosion' for the good of life on Earth, but it is clearly not good for the health of the human species. Obesity! Cancer! Diabetes! Sickness! Suffering! Anguish!

There is another aspect to the immense suffering that the master modifier inevitably brings to itself. This is the suffering/harm/ill-health that arises from 'engulfment in technology'. As the age of the 'technological explosion' matures the human species becomes literally engulfed in technology, surrounded by technology. Technology comes to pervade the everyday life of individual humans. The point comes when a typical ecstatic thing, a typical human, will be purposefully interacting with technology for almost 100% of its waking life! We are already well on the way to this point. In many countries in the world, in the year 2021, the vast majority of children spend most of their waking hours not playing in nature, but purposefully interacting with technological creations. You will see these children wandering the streets with headphones on, or with mysterious plugs in their ears which channel sounds from an extremely advanced technological device which is in their pocket or held in their hands. You will see these children using mobile phones to do a plethora of things such as play games, take photos, and communicate with other ecstatic things. You will see these children using laptop computers and gaming devices. You will see teenagers speeding around on bicycles, on motorbikes and in cars. It is extremely rare to see children who are wholly free of technology, children who are not purposefully interacting with technology and who are looking at flowers and trees, rather than the screens of technological devices. They seem to find such screens much more interesting than flowers! It is not just children; much of this already applies to large parts of the rest of the human species.

Let us explore another aspect of 'engulfment in technology'. A major cause of mental illness in both children, and an increasing proportion of ecstasy, arises from the internet, and in particular the phenomenon known as 'social media'. This phenomenon involves ecstatic things creating a personal profile on a website, from which they provide a commentary on their life through the posting of photos and updates; through such postings immense efforts are often made to convince others that one has an exceptionally interesting and happy life. Technological devices are also used to facilitate communication via social media applications. As the 'technological explosion' reaches its climax these activities come to dominate the lives of ecstatic things. This aspect of 'engulfment in technology' brings great mental suffering – it invites comparisons between people, creates pressure to 'look good', is a fertile breeding ground for hate speech and abuse, it generates anxiety, lowers self-esteem and increases feelings of depression. Despite being called 'social media' it also leads to loneliness, as real meaningful face-to-face connections between people are increasingly abandoned and replaced by degraded superficial communication via tiny screens. High levels of social media and internet use also affect physical health through lowering sleep quality; this is due to the light emissions from the screens of technological devices disrupting melatonin production.

This 'engulfment in technology' occurs because it is part of the unfolding of the journey that is the universal pursuit of ecstasy; that is to say, it occurs because the humans that are bringing forth and purposefully interacting with the technological devices experience 'increasingly intense pleasantness' through such activities. However, as we have seen, all is not rosy, because this engulfment brings immense suffering and ill-health. That which is engulfed in technology becomes physically and mentally ill. We earlier noted, when discussing the partial severance of contemporary humans from their inner qualitative wisdom, that we are currently living at a time which is marked by phenomena such as an obesity crisis, a surge in mental illness, and a rapid increase in the prevalence of cancer. It is not surprising that these phenomena are concentrated in those parts of the human species that have the highest levels of 'engulfment in technology'. We now have more of an idea why these phenomena exist. Being engulfed in technology, the master modifier becomes disconnected from that which is conducive to its own health. In order to make its host planet a more healthy and life-sustaining place, the human species itself has to suffer immensely.

Another way of describing this disconnection is to say that the human species becomes partially severed from its inner qualitative wisdom in the age of the 'technological explosion'. Whilst the master modifier acts in a way that raises itself to 'increasingly intense pleasantness' / higher states of ecstasy, its deluded state, due to its engulfment in technology, means that whilst such actions are good/desirable, a state of progression in the universal pursuit of ecstasy, enabling the human species to fulfil its cosmic purpose, they simultaneously bring immense suffering and ill health to the human species. In the age of the 'technological explosion' the master modifier is a 'cosmic puppet', acting in a way which it believes to be in its best interests, and which ultimately is in its / life on Earth's best interests, but which involves it suffering immensely for the benefit of life on Earth. So, to talk of the 'partial severance' of humans from their inner qualitative wisdom is to realise that humans are acting in a way which is ultimately not in their own best interests as individuals, whilst simultaneously being in the interests of life on Earth. This 'partial severance' ultimately gets healed through the phenomenon of spiritual development. This means that during the age of the 'spiritual explosion' the 'partial severance' will gradually close.

The suffering in the human species, suffering caused by human prejudice, by individualistic divisions, by technological accidents, by the malicious use of technological devices, by poor diet, and by 'engulfment in technology', increases until the age of the 'technological explosion' reaches its pinnacle. We are currently living at a time in which the 'technological explosion' is still in top gear, it still has a long way to go until it reaches its pinnacle. This means that the suffering and mental anguish within the human species is set to continue to increase in the immediate future. Oh, master modifier, how you suffer! Let us look forward to the age of the 'spiritual explosion' when this suffering will be reduced to a bearable level!

The age of the 'technological explosion' is brutal. There is no getting away from this fact. The master modifier, in its deluded state, causes immense suffering to its non-ecstatic planetary companions, and it inflicts immensely more suffering, a colossal amount of suffering, onto itself. The age of the 'technological explosion' could quite aptly be called 'the age of suffering' or 'the age of separation', because whilst neither suffering nor separation are new phenomena in this age, they are both immensely magnified and they both reach their pinnacle as the bringing forth of technology reaches its pinnacle.

Tool use, advanced tool use & technology

The universal pursuit of ecstasy is a journey of the increasing modification of the other, which at its climax ultimately transforms into the increasing modification of itself. Atoms are modifying the other when they come together to form molecules, molecules are modifying the other when they come together to form cells, cells modify the other when they metabolise, and humans modify the other when they construct a telescope. Part of this journey is the transition from tool use, to advanced tool use, to technology. In one sense these labels are artificial, because the journey of increasing modification that is the universal pursuit of ecstasy needs to be seen as one continuous process of unfolding. In another sense these labels are useful, because they help us to make sense of the processes that have resulted in the bringing forth of ecstasy, and they help us to make sense of how ecstasy differs from the non-ecstatic life-forms of the Earth.

Tool use on the Earth is something that long preceded the bringing forth of ecstasy and its immediate ancestors. Many diverse life-forms on the Earth mastered the art of using tools, such as twigs and stones, long before the human species came into existence. Our ancestors, the hunter-gatherers, were users of tools. The hunter-gatherers started off as just another tool using life-form, one of numerous such life-forms on the Earth. However, they became something unique; they started to bring forth tools that were much more complex than any other life-form on the Earth could create. At this point tool use had progressed into advanced tool use, and such advanced tool use was the sole preserve of one type of life-form, our non-ecstatic 'non-human biological human' ancestors. The tools used by hunter-gatherers – bow and arrow, harpoon, atlatl, projectile points – are advanced tools.

The age of advanced tool use is a long one. The advanced tools that we have just identified – bow and arrow, harpoon, atlatl, projectile points – are simple advanced tools. Advanced tools can be much more complex than this. The family of advanced tools is broad and wide. After the first advanced tools came into existence they complexified into increasingly advanced forms, as the bringers forth of these advanced forms became more highly adept at modifying their surroundings. At some point, advanced tools turned into technological creations. This point is an event of momentous importance; it is the coming into existence of ecstasy, the master modifier of its solar system! We have already considered the question of when the exact point was that the bringing forth of advanced tools morphed into the bringing forth of technological creations, and we concluded that technological creations are things which have been designed with knowledge gleaned from science. So, the bringing into existence of the first advanced tool that was designed with knowledge gleaned from science, was not the bringing forth of an advanced tool; it was the bringing forth of technology!

The environmental destruction caused by the technological birthing process

When life on a planet brings forth technology into existence it has initiated a birthing process. Technology slowly seeps across the planet at first. Then, as momentum is built up, there comes a point when the speed of the creation of new technologies increases phenomenally and the number of purposeful interactions with technological devices accelerates along a mind-bogglingly exponential trajectory. When this occurs the planet has entered the era of the 'technological explosion'. This explosion is the heart of the technological birthing process which involves life on a planet bringing forth a protective shield, a set of technological armour, in order to help boost its survival chances.

This technological birthing process inevitably brings widespread and dramatic changes to its host planet; these changes are so immense that humans have coined a phrase to refer to them: 'environmental destruction'. Let us consider some of these changes: rainforests have fallen to the chainsaw and mechanised agriculture; concrete jungles have replaced vegetative habitats; human movements around the world facilitated by airplanes and ships have spread 'alien' species and diseases which have caused the decimation of 'native' species; oil spills have decimated oceanic life in particular regions; the oceans have been plundered of their life-forms by human fishing technologies; the technological release of underground stored carbon is affecting the global atmosphere. The use of technology has directly caused these enormous changes.

In addition to this direct route, the technological birthing process has caused immense environmental destruction through an indirect route. It has led to the escalation of the human population size due to scientific/technological knowledge providing a super-advanced ability to make modifications that enable ecstasy to survive and thrive; for example, advanced medical equipment, a plethora of pharmaceuticals, devices that enable increased food output, and the creation of protective/healthy living spaces. The technological birthing process also involves an escalating increase in the per capita consumption of the Earth's resources, as ecstasy keenly interacts with the new technological fruits that have been brought into existence. So, the technological birthing process involves an escalating number of humans who are on average using increasingly more resources.

Given the magnitude of these direct and indirect technology-induced changes, it is no surprise that the phrase 'environmental destruction' has been coined to refer to them. Planetary biodiversity has been affected to such an extent that there is widespread talk of a human-induced mass extinction of life on the planet.

Why is the Universe the way that it is?

Let us pause for a moment and consider this question: Why is the Universe the way that it is? What a question! It needs to be addressed, but it cannot be answered! An account of 'the way that things are' is an account of the nature of the Universe. Such knowledge can be attained by the human species. It is possible to know the nature of the things that presently exist. It is possible to know why things have evolved and unfolded the way that they have in the past. It is possible to know how things are set to evolve and unfold in the future. This is impressive enough in itself. If one comes to know the universal pursuit of ecstasy, then one will have knowledge of all of these things.

However, one might want to go further. One might want to know why the universal pursuit of ecstasy exists. Why is the Universe this way, rather than another way? For example, why isn't the Universe comprised of things which are wholly devoid of qualitative feeling? There comes a point when the questions need to stop; they need to stop because there are no answers that a human will ever be able to give. There is not really any point pondering such questions. One just needs to accept that the Universe is the way that it is.

Could the Universe have been completely different? Could the universal pursuit of ecstasy not exist? Stop! Please stop! Just appreciate the wonderful Universe for what it is; get to know it. Stop letting your rationality go into overdrive; there is no need.

The cosmic significance of technology

Why is ecstasy – the act of bringing technology into being and purposefully interacting with that technology – the zenith of the evolutionary progression of a solar system? We don't need to say that this is just part of 'the way that things are'; we can make a lot of rational sense about why this should be so. Firstly, we can make the not too outlandish assertion that the things in the Universe are divided into living things and non-living things. Secondly, we can easily reach the rational conclusion that those things that are living are more precious and valuable than those things that are not living; who could reasonably believe the opposite, that non-living things are more precious and valuable than living things?! Thirdly, we can come to fully appreciate that life requires particular conditions in order to survive and thrive, and that these conditions will never exist in the overwhelming majority of our Solar System. This is because the unfolding of our Solar System entails that there is only a relatively short window within which life can first evolve and then establish itself so that it can maintain itself. This window is both a spatial window – the Earth is the only place in our Solar System where life can both evolve and establish itself, and a temporal window – there is only a limited time period within the swirling cycles of Solar-Systic unfoldment within which the conditions on the Earth are favourable for life to evolve and establish itself, which it does through moving through the initial living stages of the universal pursuit of ecstasy. Fourthly, we can come to see that life has been both complexifying and spreading out across the Earth; furthermore, we can appreciate that it has done this in order to bolster its ability to modify its surroundings so that it can maintain the conditions that are favourable for its continued existence and thereby increase its chance of staying in existence. In other words, we can come to appreciate that life wants to survive and that life strives to survive. Fifthly, we can draw the seemingly obvious conclusion that the bringing forth of ecstasy, the master modifier, maximises the likelihood that life will continue to survive in the future. This is because a greater modification ability directly equates to an increased probability of survival. A greater modification ability entails a movement along the scale from being controlled by external factors, to controlling these factors in order to make them more favourable for continued existence. So, it is exceedingly easy to understand why technology should be striven for by life, to appreciate why ecstasy is the zenith of the evolutionary progression of life on Earth. There is nothing here that violently clashes with that which is easily rationally comprehensible.

In short, the reason that technology is beneficial is that it increases the probability that life on Earth will stay in existence; this means that the bringing forth of technology is in the interests of planetary life. A life-bearing planet that brings forth technology is much more likely to remain a hospitable home for life than a life-bearing planet that is devoid of technology. A solar system that brings forth technology is much more likely to remain a hospitable home for life than a solar system that is devoid of technology. Let us rejoice in the fact that the Earth has evolved to the point where it has brought forth technology! The Earth is currently buzzing with ecstasy! All is well on the Earth!

This rational case as to why ecstasy is the zenith of the evolutionary progression of life on Earth is simplicity itself. However, coming to fully appreciate how this outcome comes about is far from simple. How does life evolve and unfold so as to ensure that it brings forth ecstasy? The intellect might want to hold on to the possibility that the evolution of life on Earth could have been very different and that the bringing forth of technology through *ecstasy = the human species = the master modifier* might never have happened. This is where an appreciation of the universal pursuit of ecstasy, an appreciation of 'the way that things are', comes into play. Such an appreciation helps rationality to comprehend how that which is obviously beneficial was always going to be brought forth into the realm of existence.

The greater the ability that life on Earth has to control, to modify, its surroundings, the greater is the likelihood that it will survive and thrive. The universal pursuit of ecstasy is a journey that involves an increasing ability to control; this increasing ability to control can also be pointed to by the words 'enhanced modification ability' – a greater ability for a thing to be able to modify/control its surroundings. The particular outcome automatically/instinctively sought by life is to stay in existence, which is an outcome that is best achieved through having the greatest possible modification ability. As a greater modification ability equates to a greater state of excitation/exhilaration, this means that life achieves its desired outcome through its instinctive desire to attain and hold onto 'increasingly intense pleasantness'. The utilisation of technology is the state of maximal modification ability, which means that technology massively bolsters the survival chances of the planetary life that brings it into existence.

Let us explore in a little more depth why it is that a greater modification ability involves more intense excitation/exhilaration. These two things are a simultaneous occurrence because an enhanced modification ability gives a thing more control over its existence, it enables a thing to interact with its surroundings in more intricate ways, and it feels good to be able to control; so, when a thing exerts increasingly intricate control over its surroundings this inevitably involves an increasing level of excitement/exhilaration. Of course, this relationship pervades the entire Universe; it pervades 'the hierarchy of spatio-temporal embeddedness'. In the realm of the human species, the master modifier, we can easily appreciate why the maximal modification ability that is the bringing forth and purposeful utilisation of technology is simultaneously the maximal state of feeling, the state of ecstasy. Technological devices enable humans to fulfil their potential, to unleash their creativity, to live their life to the fullest, to thrive and excel. Given this, it is easy to appreciate why the maximal modification ability that is the bringing forth and purposeful utilisation of technology is accompanied by feeling states of ecstasy. If there were no cameras, no pianos, no televisions, no computers, no airplanes, no cars, no radios, no trains, no phones, no high-end tennis rackets, imagine how dreary human life would be; human creativity and expression would be truly stifled and the ecstatic feelings that could have been experienced by our Solar System wouldn't come into existence.

Let us further consider why an enhanced modification ability is desirable through a simplistic scenario involving 'trees' and 'mammals'. A 'tree' is a thing which is at a lower stage of the universal pursuit of ecstasy than a 'mammal'. This means that a 'tree' has a lower modification ability than a 'mammal'. It follows that if the only living things to exist in a particular location were 'trees', then the life that exists in this location would be less likely to survive than if that life was constituted solely out of 'mammals'. If the conditions in this particular location were to suddenly become unfavourable for the life that is located there, the 'trees' would all die, but the 'mammals' could quickly move to another location! So, if life transforms itself from a state of 100% 'treeness', to a state of 50% 'treeness' and 50% 'mammalness', then life will have bolstered its survival chances through having an enhanced modification ability. Modification ability / control is an interaction between a thing and other things (the things that surround it and affect it); a 'mammal' is a thing which has greater control over its interactions with other things than has the thing that is a 'tree'. In every thing there is a balance between the degree to which the thing is controlled by its surroundings and the degree to which it controls those surroundings. The universal pursuit of ecstasy is a journey which entails a decrease in the former and an increasing movement towards to an increase in the latter.

Asteroids, supervolcanoes & alien invasions

It is time for us to explore the various ways in which technology bolsters the survival chances of the life that brings it into being. One of the most obvious ways in which technology bolsters the survival chances of life on Earth is through guarding against the threat of potential annihilation from a massive asteroid collision with the Earth. A small asteroid strike wouldn't threaten the existence of planetary life, it would just cause death and suffering to some living things, to some parts of planetary life (appalling death and suffering which could be prevented if the appropriate 'asteroid collision prevention technology' existed). However, a sufficiently large asteroid strike would have enough power to threaten planetary life in its totality. In this scenario the universal pursuit of ecstasy would be thwarted; the life that has arisen in our Solar System would be decimated and ultimately be extinguished. How tragic this would be! Whilst it is unlikely that a sufficiently massive asteroid strike would instantly wipe out all life on Earth, the only thing that would survive would be 'non-complex life' – extremely simple things such as bacteria, and these are things which have an exceptionally low level of excitement/exhilaration compared to that which currently exists. The strike would have transformed the Earth from a buzzing place which is the home of ecstasy and various other high levels of excitement/exhilaration, to a dull place where there is just a very very low level of excitement/exhilaration. The planet would have been knocked for six, planetary life would have been decimated; furthermore, it would be powerless to move along the journey that is the universal pursuit of ecstasy because of the passing of the Solar-Systic temporal window within which life can evolve, complexify and thrive. The Earth might not be totally lifeless, but it would effectively be home to the living dead. All would be lost! Ecstasy would never be able to evolve again in our Solar System.

Thankfully, technological development programmes are already well under way which have the objective of deflecting or destroying asteroids which are on a collision course with the Earth. Two such programmes are NASA's Near Earth Object Program and the European Union funded NEOShield project. There are a vast array of technological collision avoidance strategies that are being considered – nuclear explosive devices, kinetic impactors, asteroid gravitational tractors and ion beam shepherds, to name but a few. What strange terms we have invented to refer to these magnificent technological protectors of life in our Solar System! Through bringing forth such technologies planetary life is bolstering its defences and doing everything that it possibly can to help ensure its future survival.

Another possible threat to planetary life is a gigantic eruption from a massive supervolcano. There are several supervolcanoes on the Earth and it is a question of when the next eruption will occur, rather than a question of if a future eruption will occur. It is plausible that such an eruption could be massive enough to pose a threat to the continued existence of the life that has arisen on the Earth, through causing the extinction of all but the simplest of life-forms. Thankfully, technology can help to avert this calamitous outcome. For example, ecstasy could construct a large biome which has an artificial atmosphere; this could be located on the surface of the Earth, under the surface of the Earth, or it could even circulate in the Earth's atmosphere. This biome would be wholly self-contained and would thus be isolated from the atmosphere and workings of the rest of the Earth. This means that the biome wouldn't be affected by the immense disruption to the Earth caused by a massive supervolcano eruption.

This technological bringing forth, the biome, would provide a home for a multitude of life-forms, a plethora of various heightened states of excitement/exhilaration, in addition to ecstasy. This human technological creation could enable life on Earth to survive and thrive, when in its absence it would be decimated. This is simply yet another example of the general rule, that a greater ability to control/modify results in a greater ability to survive. When the external planetary environment returned to a favourable state, which might be a very long time after the eruption, the life-forms that were inside the biome would be able to leave the biome and repopulate the planet. As with a massive asteroid strike, such a survival and thriving of life after the event would not be possible in the absence of technology, for the Solar-Systic temporal window which enables the advancement of planetary life from simple beginnings to advanced life-forms has already passed.

On a slightly more fruity level, one might be tempted to believe that the bringing forth of technology bolsters the survival chances of life on Earth through giving protection against a possible alien invasion of the planet. However, in reality this is not a benefit of the bringing forth of technology. For, if life-forms have technology that is advanced enough to enable them to visit our planet, they will have inevitably passed through both the 'technological explosion' and the 'spiritual explosion' on their home planet. This means that they would be spiritually enlightened and would have come to see how things are through 'the realisation of cosmic purpose'. In turn, this means that they would see us, in our present state of development, as 'children' which need to be cherished and nurtured, rather than as 'enemies' which need to be attacked or destroyed! In other words, their objective would be to help us to advance along the journey that is the universal pursuit of ecstasy.

Whilst technology can be deployed in various ways in order to enhance the survival prospects of life on Earth, there is one particular deployment which is much more significant than all of the others. Coming to fully appreciate both the need for this particular deployment, and the urgency of this need, is our next objective. It is exceedingly easy to understand that a technological deployment that knocks a massive asteroid off its collision course with the Earth is a deployment that can save life on Earth from decimation. It is also very easy to understand that the technological deployment that is the biome is a deployment that can save life on Earth from decimation following a massive supervolcano eruption. In contrast, the most significant deployment of technology that life on Earth requires in order to survive and thrive has many more aspects, which makes it harder to fully appreciate. The deployment in question is the sufficient deployment of technology.

The sufficient deployment of technology

You might recall that when technology has sufficiently complexified and been sufficiently deployed, that life on a planet then moves to the final stage of the universal pursuit of ecstasy, which is the 'spiritual explosion'. Given the magnitude of the suffering that exists in the age of the 'technological explosion', you might be keen to know what it means to talk of the sufficient deployment of technology. For, once this deployment has occurred we will then be speeding at full pelt into the 'spiritual explosion', where suffering will be magnificently reduced. We have just been exploring the cosmic significance of technology, the various ways in which the bringing forth of technology enhances the survival chances of life, and we noted that one particular deployment of technology is the most significant. It is this particular deployment of technology, the use of technology to actively regulate the temperature of the Earth's atmosphere, which is the sufficient deployment of technology. When technology has been sufficiently deployed, when it has been deployed to actively regulate the temperature of the

Earth's atmosphere, we will then have reached the stage where the 'technological explosion' is able to be superseded by the 'spiritual explosion'. There is clearly much which needs to be said before one can fully comprehend the importance of the sufficient deployment of technology.

The need for this deployment is exceptionally easy to comprehend. We have already come to appreciate that the Sun's output – the amount of solar radiation that it propels outwards throughout our Solar System – increases as our Solar System unfolds/evolves/ages. This increasing propelling affects the way that things unfold throughout our Solar System. This increasing propelling is the reason why there is only a small temporal window within our Solar System within which life can evolve, establish itself and reach ecstasy. This increasing propelling is the reason why the Earth is the womb of Solar-Systic life. The problem is that in an unfolding solar system – a solar system characterised by forever increasing levels of solar radiation – it is inevitable that that which is the bringer of life, at a certain point becomes the enemy of life. This is one thing that is more certain than 'death and taxes'. There is nothing more certain!

Life needs particular conditions in order to survive and thrive; in particular it needs an atmospheric temperature that is neither too hot nor too cold. The forever increasing output of the Sun as our Solar System unfolds, the warming of our Solar System, inevitably means that the atmospheric temperature of the Earth is on course to become too hot for planetary life to thrive, and ultimately to survive. In other words, the increasing output of the Sun becomes the archenemy of the universal pursuit of ecstasy! What can planetary life do in the face of such a powerful and unrelenting enemy? It can complexify, increase its ability to modify, reach ecstasy, and deploy technology; it can deploy technology in order to offset the warming of our Solar System and thereby maintain the atmospheric temperature range on the Earth that life needs in order to thrive. Of course, such a deployment is the sufficient deployment of technology.

The need for this technological deployment is exceedingly obvious; there is an overwhelming scientific consensus both that the Sun's output will forever increase as our Solar System ages, and that life needs a particular atmospheric temperature range in order to survive and thrive. The only conclusion that can follow from these two realisations is that the atmospheric temperature of the Earth is set to increase to a level that is too hot for life to thrive (ecstatic life to exist and states of highly excited/exhilarated life to exist) and ultimately to survive. However, it is possible that the atmospheric temperature can, for an exceptionally long stretch of time, be prevented from rising through the sufficient deployment of technology – the deployment of technology to actively regulate the Earth's atmospheric temperature. It is the purpose of ecstasy, the purpose of the human species, to prevent the atmospheric temperature from rising (due to the increasing output of the Sun) through bringing into being the sufficient deployment of technology.

All of this might be blindingly obvious. However, the really important question is when in the future such a technological deployment will be required. The answer to this question is of paramount importance, yet it is not immediately obvious. It is easy to believe that the need for the sufficient deployment of technology is way in the future, thousands of years, hundreds of thousands of years, even millions of years. Such a belief is wrong! Our mission is to clearly understand why this belief is wrong, to understand why there is an imminent need for the sufficient deployment of technology. In other words, our mission is to come to appreciate why the survival of ecstasy, and the survival of complex life on Earth, requires the deployment of technology to actively regulate the Earth's atmospheric temperature in the very near future.

The sufficient deployment of technology involves ecstasy deploying technological devices so that the Earth's atmospheric temperature can be prevented from rising to a level that is too hot for the continued thriving of planetary life. When we have come to clearly appreciate why such a deployment is imminently required, we will then be in a position to understand the nature of the technological devices that need to be deployed. We will come to see that basic technological atmospheric temperature regulation – the deployment of devices which modulate 'greenhouse gases' – is not powerful enough to be the sufficient deployment of technology.

The trap of the flawed extrapolation

When one considers the relationship between ecstasy and the rest of life on Earth it is easy to fall into a trap. It is easy to believe that what has happened in the past will also happen in the future. Life on Earth has been positively thriving for millions of years without technology, so it is easy to fall into the trap of believing that if the human species were to suddenly, in the year 2021, vanish from our Solar System, that life on Earth would continue to thrive. If one makes this simplistic extrapolation from the past to the future then one has made a fundamentally flawed extrapolation.

Life on Earth has, in the past, gone through multiple episodes of mass extinction which have involved a high proportion of the variants of life that have inhabited the planet ceasing to exist. After all of these episodes life on Earth recovered, it bounced back to a level of health that superseded that which existed prior to the devastating mass extinction. It is easy to fall into the trap of believing that because this has happened in the past that it will, in the absence of ecstasy, also happen in the future. If one makes this simplistic extrapolation from the past to the future then one has fallen into the trap of the flawed extrapolation.

This flawed extrapolation, this way of thinking, arises because of the failure to appreciate the subtle and interconnected way in which our Solar System is an evolving and unfolding entity. Simplistic extrapolations do not sit well within the context of an evolving and unfolding entity. It is because we exist that these extrapolations are flawed. The fact that our Solar System has unfolded to the point that has enabled the bringing forth of ecstasy, its master modifier saviour, means that its womb of life, the Earth, is in a very delicate state of balance in its ability to continue to support life. The bringing forth of ecstasy means that life on Earth's 'first line of defence' (which we will explore shortly) against the global warming of our Solar System is close to failure; this means that life on Earth's ability to bounce back and thrive is under immense strain. The cause of this strain is obviously not ecstasy; it is the age of our Solar

System, an age which is of such a long duration that it has enabled ecstasy to evolve. At this stage in our unfolding Solar System the past is no longer a reliable guide to the future when it comes to the ability of planetary life to bounce back from a period of weakness (a mass extinction episode). In other words, the very fact that we exist means that if we extrapolate from the past we will fall into the trap of the flawed extrapolation.

The risks of technology to life on Earth

Before we delve into our exploration of the imminent need for ecstasy to deploy technology which actively regulates the temperature of the Earth's atmosphere, it is perhaps useful to say a few words concerning the risks of technology. After all, you might instinctively be a 'techno-sceptic' who believes that technology poses a possible threat to the future existence of life on Earth. In its most extreme form, techno-scepticism entails the belief that life on Earth would be better off without technology. This is the erroneous belief of the deeply deluded. Let us be clear, to talk of the risks of technology to life on Earth is to talk of technology as a potential threat to the continued existence and/or thriving of life on Earth; it is not to talk of technology as a risk to the continued existence of a particular individual life-form, or a collection of individual life-forms. We have already considered this; we have explored the immense death and suffering that ecstasy, the bringer forth of technology, the master modifier, brings into existence both for itself and for other particular life-forms on the Earth. So, all is clearly not rosy in the technological garden. However, to believe that technology poses a threat to the continued thriving/existence of life on Earth, is a very different thing from acknowledging that the bringing forth of technology entails immense suffering and death for particular individual life-forms.

How could technology possibly pose a threat to the continued existence of life on Earth? One possibility that is often mooted is that technological things could be created which feed on living things. If these technological entities could replicate themselves, then they could spread out across the planet in an immensely short period of time; they could kill every living thing in their path. Don't worry about this actually happening in the future; it is a rather fanciful futuristic speculation. Another possible technological doomsday scenario is a nuclear apocalypse. Could a nuclear war wipe out ecstasy? Could a nuclear war stop life on Earth from thriving? Whilst a nuclear war would undoubtedly cause immense suffering on a massive scale, it wouldn't wipe out ecstasy and other complex life-forms from the face of the Earth. And, to repeat, causing suffering and death to particular individual life-forms, and being a risk to life on Earth, are two very different things.

The most plausible threat from technology to the continued existence and thriving of life on Earth is rather more mundane. No self-replicating feeding-on-life entities, no nuclear apocalypse, simply the everyday use of technology across the planet, the effects of which build up over hundreds of years to an aggregate effect which is of immense significance. We are here talking about the aggregate effect of technological lifestyles slowly but surely having an impact on the functioning of the Earth's biosphere. The cycles of the Earth's biosphere are often referred to as biogeochemical cycles, and they determine things such as the temperature of the Earth's atmosphere. The human population, ecstasy, is currently very large; whilst, technological devices are, by their very nature, devices of master modification. Think of the transformations made to the Earth by technological devices such as the chainsaw, the car, the airplane and the ship! Given the large size of the human population and the nature of technological devices, it is entirely plausible that ecstasy could have an aggregate effect over time which leads to a massive change, a 'transition shift', in the way that the biogeochemical cycles of the Earth operate. This could entail a sudden massive change in conditions which make the Earth inhospitable for ecstasy and for other highly excited/exhilarated life-forms. In short, it is plausible that ecstasy could destroy itself and decimate life on Earth through its everyday use of technology.

Is the scenario outlined in the previous paragraph going to occur? Of course it isn't! What would be the point of the universal pursuit of ecstasy, a pursuit for the ability of master modification, if the master modifier was going to destroy itself through its ability to master modify? The Universe isn't that ridiculous! The Universe isn't that pointless! In order to clearly see why this isn't going to occur we need to pick up where we left off a little earlier; we need to explore why there is a need for ecstasy to deploy, in the near future, technology which actively regulates the temperature of the Earth's atmosphere. For, the cumulative effects of everyday technology use, and the imminent need for ecstasy to deploy technology which actively regulates the temperature of the Earth's atmosphere, are tightly interconnected aspects of the unfolding journey of our Solar System that is the universal pursuit of ecstasy.

The warming of our Solar System & the cosmic game

We are currently in the process of exploring why the master modifier needs to deploy technology to actively regulate the temperature of the Earth's atmosphere in the near future. As we move forward with this exploration we need to always keep in mind that we live in a warming Solar System. Our Solar System is an evolving and ageing whole. With every cycle of swirling that the planets make around the Sun, our Solar System ages a little more; our Solar System unfolds and evolves a little more. Everything is in a constant state of evolving and becoming. The driver of the ageing of our Solar System is the increasing output of the Sun, which causes our Solar System to progressively warm until it becomes too hot to exist! There is obviously another major driver in our Solar System: the universal pursuit of ecstasy. There is a delicate interplay between these two drivers. The strivings of the universal pursuit of ecstasy always need to be seen against the backdrop of the increasing level of solar radiation which is ageing our Solar System. Another way of putting this is to say that the universal pursuit of ecstasy is affected by the increasing level of solar radiation which ages our Solar System; it is either helped or hindered by this increasing.

As we have already explored, the Earth is the womb of Solar-Systic life; it is the only place in our Solar System where life can thrive, ultimately reaching its ecstatic zenith. If we go back a very long way in time, to the immediate aftermath of the Earth's formation, planetary conditions were not suitable for the universal pursuit of ecstasy to flourish; this is because the atmospheric temperature was very hot. It is important to appreciate that this extreme hotness was not due to incoming solar radiation; rather, it was due to the explosive processes of planetary formation. At this time the Sun was the friend of planetary life, because the amount of solar radiation that it radiated throughout our Solar System was relatively low. When the Earth's atmospheric temperature cooled to reasonable levels, following the extreme highs caused by planetary formation, the relatively low level of incoming solar radiation to the Earth enabled life to evolve. At this time the incoming level of solar radiation was on the low side of what is ideal for the thriving of life. As time passed, as the level of solar radiation increased, conditions became perfect for life to develop and thrive through the universal pursuit of ecstasy.

As our Solar System continued to age, as the amount of solar radiation that reached the Earth increased, life on Earth was easily able to deal with this and keep its atmospheric temperature favourable for planetary life to flourish. So, life on Earth was able to power forward along the journey that is the universal pursuit of ecstasy. However, things had to change. Given that life cannot flourish in an atmosphere that is too hot, the time inevitably comes when the Sun becomes the enemy of the universal pursuit of ecstasy. In other words, the time comes when life on Earth is no longer not-hindered by the increasing level of solar radiation; quite the opposite, its striving to stay in existence becomes deeply hindered by the increasing solar radiation. The Sun became the enemy of the universal pursuit of ecstasy a long time ago. The poor universal pursuit of ecstasy! How can you defeat such a mighty and unrelenting enemy?

We can imagine that a cosmic game is in full swing. Can the universal pursuit of ecstasy reach the sufficient deployment of technology before the increasing level of solar radiation stops the universal pursuit of ecstasy dead in its tracks? This is the all-important question. This is the game. We need to always keep this cosmic game in the back of our minds because it reminds us that if the life that has arisen on the Earth, the life that so desperately wants to survive, is to have a long-term future, then it needs to reach the sufficient deployment of technology. Technology is a necessity. There is no room for doubt. There are no questions to be raised concerning this. It is just a simple fact. You might have a personal dislike of technology, and seek to live a technology-free life, which is fine, but there is no getting away from this simple fact. Planetary life needs to bring forth technology which actively regulates the temperature of the Earth's atmosphere if it is to survive. Whilst keeping in mind this cosmic game, and the need for the sufficient deployment of technology which is an integral part of it, we can return to the question of when this deployment is required. Are we talking tens of years, hundreds of years, thousands of years, or hundreds of thousands of years?

The ongoing war between life on Earth & the Sun

The first thing that we need to clearly envision is how the war between life on Earth and the Sun has so far played out. One cannot come to appreciate what will happen in the future unless one has knowledge of the factors that have caused the present to be as it is. This is particularly the case in terms of the war that we are considering here, because that which has happened in the past is still having impacts in the present and it will have its most potent impacts in the future.

A central player in the war is the phenomenon that humans call the 'greenhouse effect'. The Sun propels solar radiation throughout our Solar System, so some of this inevitably hits the surface of the Earth. After it hits the Earth's surface some of the solar radiation gets absorbed by the surface and the rest bounces off the surface into the Earth's atmosphere. The atmosphere of the Earth contains some molecules which have properties that cause the 'trapping' of some of this 'bouncing' solar radiation; these molecules are commonly referred to as 'greenhouse gases'. So, whilst some of the 'bouncing' solar radiation heads straight back out into the wider Solar System, the rest gets intercepted by the greenhouse gas molecules in the Earth's atmosphere. The effect of this 'trapping' by the greenhouse gas molecules in the Earth's atmosphere is to cause the temperature of the Earth's atmosphere to be higher than it otherwise would be. This is the 'greenhouse effect'. An increase in the number of greenhouse gas molecules in the atmosphere enhances the 'greenhouse effect' and thereby results in a higher atmospheric temperature.

As the war between life on Earth and the Sun has progressed, the greenhouse gases in the Earth's atmosphere have played a pivotal role. The key greenhouse gas has been carbon dioxide. Before the war started, at a time when the Sun was not a hindrance to life on Earth, the carbon dioxide concentrations in the Earth's atmosphere were at a very high level. At this time, conditions were fine for life to survive and thrive and the universal pursuit of ecstasy was in full swing, with life complexifying and increasing its modification ability. As time progressed, as the planets continued to swirl, as the level of solar radiation continued to increase, there was an increasing pressure being exerted on the Earth's atmosphere to become hotter. But life on Earth didn't want the Earth's atmosphere to become hotter! Life already had the atmospheric temperature that it needed to thrive; it didn't want the atmosphere to become hotter! Bad Sun! The Sun had become the enemy of planetary life. Thankfully, at this time in the unfolding of our Solar System the Sun was a rather timid and weak enemy. Life on Earth could easily deal with the increased pressure being exerted for a higher and unwelcome atmospheric temperature. All life on Earth had to do was to utilise its 'greenhouse effect'. Through reducing the amount of carbon dioxide in the Earth's atmosphere, life on Earth offset the pressure exerted by the Sun for a higher atmospheric temperature and thereby maintained the atmospheric temperature that it needed in order to thrive. In other words, the amount of incoming solar radiation had increased, but this was offset by more of that which bounced off the surface of the Earth going straight back out into the wider Solar System, due to there being fewer greenhouse gases in the Earth's atmosphere.

The first line of defence

How did life on Earth achieve this impressive feat? How did it win its initial battles against the Sun? Life on Earth pulled carbon dioxide out of its atmosphere and absorbed it into itself via both oceanic and land-based life-forms. When these life-forms died, the absorbed carbon gradually got buried and stored under the surface of the Earth. This process of utilising its 'greenhouse effect' is the first line of defence which life on Earth deployed in its war against the Sun. It has been a sturdy and long-lasting defence, because for an immensely long time life on Earth has been able to respond to the increasing pressure being exerted by the Sun for a higher atmospheric temperature by pulling an increasing amount of carbon dioxide out of its atmosphere. The first line of defence has given life on Earth lots of time, precious time, to continue the journey that is the universal pursuit of ecstasy. The first line of defence has been able to be so successful because when the war started there was an enormous amount of carbon dioxide in the atmosphere that could be moved to underground storage.

The first line of defence has done a tremendous job. However, there was, of course, always a glaringly obvious issue looming on the horizon. This mechanism utilised by life on Earth could not go on forever; it is strictly time-limited; its effectiveness ebbs away as our Solar System ages. After all, the amount of carbon dioxide in the Earth's atmosphere is not unlimited! This means that it is inevitable that the first line of defence will become overwhelmed and expire. The lower the level of carbon dioxide concentrations in the Earth's atmosphere, the harder it becomes for life on Earth to mobilise its first line of defence. Carbon dioxide concentrations in the atmosphere can only fall to a certain level, so the first line of defence ultimately has to breakdown and expire. Can you envision the breakdown of the first line of defence? Life on Earth struggles to flourish as atmospheric carbon dioxide concentrations decline past a certain point; there is a long period of struggle, of weakening of the first line of defence, which is followed by its collapse. The gradual breakdown of the first line of defence has resulted in tumultuous fluctuations on the Earth as it has spluttered between periods that we call 'Ice Ages' and 'Interglacial Periods'. These fluctuations are indicators of planetary sickness; they are a sign that life on Earth is struggling to keep its atmospheric temperature down to a level that is favourable for its continued existence. However, this weakening and inevitable collapse isn't really a failure. After all, the purpose of the first line of defence was only ever to buy time, time for life on Earth to move along the journey that is the universal pursuit of ecstasy and bring forth its main line of defence, the ecstatic fruit that is the sufficient deployment of technology. The bringing into existence of ecstasy heralds the bringing into existence of the main line of defence which life on Earth needs to prolong its war with the Sun and stay in existence.

The main line of defence & the partial reversal of the first line of defence

The bringing into existence of ecstasy, the master modifier, is an event which has lots of interlinking aspects. Some of these aspects are seemingly contradictory, some of these aspects can be hard to grasp, and ecstasy itself inevitably becomes blinded to some of these aspects out of necessity. Such a 'blinding' is needed so that ecstasy can fulfil its cosmic purpose of bringing forth life on Earth's main line of defence – the sufficient deployment of technology.

One curious aspect of the bringing forth of the main line of defence – the sufficient deployment of technology – is that it entails a partial reversal of the first line of defence. The main line of defence requires a master modifier to bring it into being, and a master modifier does what you might expect. The master modifier explores and investigates its home planet in great detail; it sees every 'thing' as a resource that it can utilise and transform, and it doesn't just see this, it acts accordingly. Investigate; utilise; explore; transform; mould; investigate more; transform more. The master modifier is at work!! In order to power its 'technological explosion' the master modifier needed to utilise the resource that was created by the first line of defence. Recall that the first line of defence involved life on Earth removing carbon dioxide from its atmosphere and transforming it into carbon which was stored under the surface of the Earth. We have created the term 'fossil fuels' to refer to some of this stored carbon. The master modifier uses its primal technologies to remove fossil fuels from under the surface of the Earth in order to power the 'technological explosion' which leads to the bringing forth of the main line of defence.

So, the act of bringing forth technology to ensure the future survival of life on Earth entails a partial reversal of that which was previously conducive to the continued survival of life on Earth. In other words, in the era of the first line of defence the war between life on Earth and the Sun entails progressively lower atmospheric carbon dioxide concentrations being favourable for life on Earth, yet the bringing forth of life on Earth's main line of defence entails temporarily raising atmospheric carbon dioxide concentrations. Atmospheric carbon dioxide concentrations used to be 5000ppm, but due to the first line of defence they were only 280ppm at the start of the Industrial Revolution. The partial reversal has nudged this figure up to just over 400ppm. That these two opposing movements – the massive fall followed by the tiny increase – are both to be welcomed, might seem to be slightly paradoxical, even contradictory, yet it is actually quite wonderful. One stage of the unfolding journey that is the universal pursuit of ecstasy has provided the resources (resources that we now call 'fossil fuels') that the next stage of the universal pursuit of ecstasy requires. How amazing is the universal pursuit of ecstasy! What is needed gets provided!

This partial reversal in atmospheric carbon dioxide concentrations is a sign that the master modifier saviour is in full swing; it is a sign that the main line of defence is being brought forth; it is a sign that all is well on the Earth! Well, all is as well as it could possibly be given the stage of unfoldment of our Solar System that we are currently living through. We are not yet into the age of the 'spiritual explosion' when things will be better. Could things possibly be any better for any of the ecstatic and non-ecstatic life-forms that are currently in existence? We will return to this question later, when our journey is nearing its conclusion.

Drawing the wrong conclusions

We have seen that the bringing into existence of its main line of defence requires life on Earth, through that part of life on Earth that is ecstasy, to initiate a partial reversal of its first line of defence. If one focuses solely on this partial reversal, if one lacks wider insight into 'the way that things are', then one could quite reasonably draw some wrong conclusions. One might draw the wrong conclusion that the partial reversal is a bad thing. One might draw the wrong conclusion that the human species is causing needless damage and disruption to the Earth. One might even draw the wrong conclusion that the human species is the 'enemy' of life on Earth rather than its precious ecstatic saviour!

At the moment, in the year 2021, there is a lot of attention given to this partial reversal, the rising level of carbon dioxide in the Earth's atmosphere. There is now a widespread realisation that this phenomenon has been caused by human activity (you might well be aware of what is typically referred to as the 'hockey stick' graph which supports this view). Such a realisation has generated panic in many quarters and has resulted in the formation of organisations such as the Intergovernmental Panel on Climate Change (IPCC) whose main objective is to stop this partial reversal. There is currently no widespread realisation that this partial reversal is part of a longer term process of unfolding which is in the interests of life on Earth. There is currently no widespread realisation of the need for the sufficient deployment of technology in the near future. If there was such a realisation, then the human species would stop worrying itself about the partial reversal and would instead focus on the bringing into being of the sufficient deployment of technology.

The partial reversal of the first line of defence is currently widely seen in the context of an overwhelmingly dominant story that has very recently been created by humans. This story proposes that the human species is fundamentally damaging the Earth due to its selfish greed, and that this damage is to the sole benefit of the human species, whilst harming non-human life. We have already seen why this story is wrong. In reality, the activities of ecstasy are for the benefit of life on Earth, and they bring much greater suffering and harm to the human species than they do to the non-ecstatic life-forms of the Earth! The purpose of our journey is to help you to see through this overwhelmingly dominant story and to come to appreciate 'the way that things are'.

It is exceptionally easy to draw the wrong conclusions. One can only draw a conclusion based on the information that one has access to. If one lives at a time which is dominated by a particular story concerning the relationship between the human species and the rest of life on Earth, then it is highly likely that the conclusions that one draws will be informed by, and will be situated within, this dominant view. The reality is that the current dominant view is wrong. The human species is the friend and saviour of life on Earth; it is not its selfish destroyer and enemy.

The two drivers of human culture

We are making good progress in our effort to understand why there is a need for the sufficient deployment of technology in the near future. In so doing we are inevitably gaining a lot of insight into 'the way that things are'. We have just been exploring how the bringing forth of the main line of defence entails a partial reversal of the first line of defence. Our next task is to gain a better insight into the way that human culture unfolds through time. After all, the partial reversal of the first line of defence is part of this unfolding. In this context, it will be fruitful to think of the unfolding of 'human culture' as encapsulating both the cultural evolution of *ecstasy = the human species* and the preceding cultural evolution which stretches back to our hunter-gatherer ancestors.

The unfoldment of human culture is, of course, driven by the universal pursuit of ecstasy, because the universal pursuit of ecstasy drives the unfoldment of all things. The universal pursuit of ecstasy is the primary driver of human culture, but this obviously isn't the whole story. As we have already explored, the universal pursuit of ecstasy cannot be considered in isolation because our Solar System is a tightly interconnected whole that goes through cycles of qualitative unfoldment as the planets swirl around the Sun and form particular alignments with each other. These cycles of qualitative unfoldment, which are associated with the swirling of the planets, are the second driver of human culture. This second driver is really a tandem effect which combines progressive increases in solar radiation emanating from the Sun, with the repetitive cycles of unfoldment of the planets around the Sun. However, when it comes to being a driver of human culture up to the year 2021, this second driver is overwhelmingly dominated by the repetitive cycles of planetary unfoldment. This is because human culture has unfolded over a relatively short period of time, a period so short that solar radiation has barely increased over this period.

These two drivers of human culture subtly interact. Of course, the universal pursuit of ecstasy does not itself operate in cycles/phases; all things are continuously and unceasingly in pursuit of ecstasy. In contrast, the cycles of unfoldment of our Solar System have effects which are cyclical. These cycles impinge on the journey that is the universal pursuit of ecstasy through providing some timeframes within which conditions are very favourable for its advancement, and by providing other timeframes within which conditions are far less conducive for its advancement. So, whilst the pursuit of ecstasy is unceasing, its advancement is cyclical, in that in some periods the pursuit speeds ahead, whilst in other periods it is very gradual or static.

The Solar-Systic cycles of unfoldment involve particular types of qualitative state becoming dominant on the Earth, and in swathes of the rest of our Solar System, in particular periods of time. These types of qualitative state impinge on the universal pursuit of ecstasy and thereby have effects on the unfoldment of human culture. So, at one moment in time the Earth will be dominated by qualitative states which, using human language, we can very vaguely point towards by using words such as: blissful/glowing/buzzing/expansive/driven/motivated. The domination of the Earth by such exuberant states provides vastly enhanced opportunities for the progression of the universal pursuit of ecstasy. Whereas, at a later time, the Earth will be dominated by qualitative states which we can loosely point towards through words such as: stifled/depressed/limiting/unfulfilled/unpleasant. This domination provides vastly reduced opportunities for the progression of the universal pursuit of ecstasy. In the former case, the Earth is powering forward along the journey that is the universal pursuit of ecstasy. This journey entails the bringing forth of life, life bringing forth ecstasy, ecstasy bringing forth the 'technological explosion' which leads to the sufficient deployment of technology, and it culminates in the bringing forth of a 'spiritual explosion' and 'the day of the realisation of cosmic purpose'. So, in this former case, human culture will be powering forward along its journey towards and through the 'technological explosion'. Whilst in the latter case, where the qualitative states that are dominating the Earth are stifling/limiting/etc., human culture will not be powering forward, it will be held back.

The two trends/forces in human cultural unfoldment

The two drivers of human culture that we have just explored – the universal pursuit of ecstasy and the cycles of unfoldment of our Solar System – operate under the radar. The trajectory of human culture is directed by these two drivers, yet their existence can be very easy to miss. In particular, the cycles of unfoldment of our Solar System, with their very different qualitative states, only become clearly visible when one is able to zoom out and see things from a wide perspective. From this wide perspective, one can see that the cycles of unfoldment of our Solar System have effects on human culture which are easily identifiable; these cycles result in two trends/trajectories/movements in human cultural unfoldment.

We need to explore how the cycles of unfoldment of our Solar System involve two opposing forces, forces which generate opposing trends in the trajectory of human cultural unfoldment. We have already considered how the Solar-Systic cycles of unfoldment bring into existence, over a particular stretch of time, a dominance of one of two types of states throughout swathes of our Solar System. The first of these two types of states can be loosely pointed to through the use of words such as blissful/glowing/buzzing/expansive/driven/motivated; whilst, the second of these types of state can be very vaguely pointed to through the use of words such as stifled/depressed/limiting/unfulfilled/unpleasant. Both of these disparate states are ever present and widespread throughout our Solar System. What the cycles of unfoldment do is to bring into existence an increasing dominance of one of these two types of qualitative state in swathes of our Solar System, and thus in human culture, over particular stretches of time. The length of these stretches of time is determined by the duration of particular planetary alignments within the overall Solar-Systic cycle of unfoldment. This process of increasing dominance is a process of increasing intensity, which is followed by a period of decreasing intensity. So, one of these types of states gradually becomes more and more pronounced until it reaches fever pitch, its period of maximum intensity, as particular planets come into full alignment; then, the intensity of this state gradually fades until it is supplanted for dominance by the other type of state, as opposing planetary alignments come into existence.

In order to see the effects of these cycles of unfoldment in human culture we need to zoom out so that we can see trends that are occurring over long periods of time. As there are two types of state within the cycles of unfoldment, we would expect there to be two identifiable trends within human culture. These trends exist both in human cultural unfoldment and in the cultural unfoldment of the 'non-human biological humans' that preceded and gave rise to human cultural unfoldment. We can refer to these two trends/forces in various ways, with various labels; for, certain labels will be more appropriate than others at different stages of cultural unfoldment. We also need to keep in mind that the trends arise from the opposing forces – forces which have a concrete qualitative existence.

In our present day, the year 2021, we can fruitfully shed some light on these two trends/forces through the labelling of them as follows: *the force to environmental destruction* and *the force to environmental sustainability*. The force to environmental destruction represents the universal pursuit of ecstasy moving forward at full pelt, through the coming into existence of states that are blissful/glowing/buzzing/expansive/driven/motivated. In contrast, the force to environmental sustainability represents the holding back, the frustrating, the reigning in, the taming, of the universal pursuit of ecstasy, through the coming into existence of states that are stifled/depressed/limiting/unfulfilled/unpleasant.

When we look at pivotal events in cultural unfoldment, events such as the scientific and technological revolutions, these events occurred at a time when the force to environmental destruction was dominant on the Earth. This force results in both periods of increased human understanding, and periods of increased human modification of the Earth. When this force is dominant intellectual and technological breakthroughs are much more likely to occur. In contrast, when the force to environmental sustainability is dominant there are periods of stagnation, where there is a lack of progress and things are being held back and reigned in. We are currently in the midst of the age of the 'technological explosion', throughout which the force to environmental destruction is dominant.

It is worth reiterating that the names of these two forces are names that are suitable given that they have been created in the year 2021. If names were being created in the year 1721 to refer to these forces, they would, no doubt, be very different. Why have these particular names been used? These names are appropriate because the dominant theme in contemporary human culture concerns the relationship between humans and the Earth viewed through the prism of 'the environment'. We are here talking about 'environmental issues'; for example: biodiversity loss, habitat transformation, the destruction of rainforests, the destruction of the ozone layer, human-induced global warming and climate change, genetically modified crops, rising sea levels and melting icecaps, recycling, sustainability, greenwashing, reducing waste, reducing energy consumption, air pollution, oceanic plastic pollution, ocean acidification, bringing forth renewable energies, and attempting to change human attitudes and behaviour to make it increasingly environmentally friendly through taxes and incentives. There is little doubt that 'the environment' is the dominant theme in human culture in the year 2021. So, the two forces within the cycles of unfoldment of our Solar System, forces which are having significant effects in human culture in the year 2021, are most easily seen and referred to through the prism of this dominant theme. Yet, we need to keep in mind that these two forces have effects which permeate human culture in its entirety.

When we look at human culture in the year 2021, when we reflect upon the way that it is unfolding, when we consider the forces that are currently in play, then we can clearly see that there are two starkly opposing forces. Firstly, there is a force which is propelling human culture towards making forever greater modifications and transformations of the Earth. Secondly, there is a force which is seeking to reign in these modifications/transformations.

The former is obviously the force to environmental destruction, whilst the latter is the force to environmental sustainability. If we zoom out and envision the state of the Earth over a few thousand years, we would see that for the first several hundred years one of these forces was dominant, that it was then supplanted for dominance by the other force for several hundred years, after which time the originally dominant force regained its dominance. Let us consider how these two vying forces are currently forging human cultural unfoldment.

The force to environmental destruction is currently the overwhelmingly dominant force in human culture. This force is the unfettered striving to modify which exists throughout the Universe and which is the hallmark of the universal pursuit of ecstasy (a greater modification ability entailing a higher level of excitation/exhilaration). This force is powered by both an escalating global human population size and the increasing amount of resources that are used by the individuals in this population. There are a number of reasons why this force is currently so powerful and overwhelmingly dominant. Firstly, humans in large parts of the world have a very high and deeply unsustainable level of resource use. Secondly, the human population size has escalated since the start of the 'technological explosion' and it is still escalating. These two factors – an escalating population, with a significant proportion of that population having a very high resource use per head – are a potent force for environmental destruction. This force is reinforced by a third factor: the vast majority of humans who have experienced living a lifestyle that entails using a high level of resources have ingrained habits / ways of living, and they are typically very reluctant to any suggestion that they should use far less resources. It would almost seem reasonable to suggest that the vast majority of these humans are incapable of making anything other than piecemeal changes to the way that they live; incapable that is, if these changes entail having less, modifying less, controlling less; such is the strength of the desire of all things to have control over and to modify their surroundings! Fourthly, this whole situation is strongly compounded by the fact that both the new members that are born into the escalating population, and the existing members of the population that are currently living on a very low resource use per head, typically, and quite understandably, aspire to attain the living standards of those who have a very high resource use per head. This combination of factors constitutes the almost unstoppable juggernaut that is the force to environmental destruction.

The second force at work in human culture is the force to environmental sustainability. This force is that part of the universal pursuit of ecstasy that contains qualitative states which have, in the journey of the universal pursuit of ecstasy up to the bringing forth of ecstasy, been wholly undesirable in terms of their qualitativeness. These are the states which have been explored and rejected. The universal pursuit of ecstasy is a striving for 'increasingly intense pleasantness'; we are here talking about unpleasant/limiting/stifled states of feeling. When this force enters the realm of human culture these qualitative states eventually come to have a positive value for the first time; these states actually become desirable! These states become intellectualised by ecstasy as being desirable, they become states which are often sought after, rather than states that are to be resisted. These states act as a necessary break on the rampant desire to modify which has resulted in the extreme maximal states of modification brought into being by the master modifier. There eventually comes a point in the unfolding of a solar system when the desire to modify needs to be reined in and constrained. One important strand of the force to environmental sustainability, as it manifests itself in human cultural unfoldment, is a penchant towards both greater reflection and heightened ethical sensitivity. There is obviously an extremely close link between this state of heightened ethical sensitivity and the guilt-tinged nature of the 'technological explosion' which we explored in this section: *Global Warming & Human Cosmic Purpose*.

The force to environmental sustainability is still in its infancy within human culture; it is a force which has only had significant strength since the second half of the twentieth century. This force includes recycling activity, the reuse of materials, the promotion of cycling, the production of electric cars, and the development of wind farms and other renewable energy sources. This force is in operation when humans campaign against deforestation, biodiversity loss and human-induced global warming, and it is in operation when political leaders meet up to discuss these issues. This force is also at work when some humans feel slightly guilty about their unsustainable lifestyles. Yet, despite this guilt, these humans continue to act in a deeply unsustainable way due to the dominance of the force to environmental destruction. For example, one effect of this guilt is to cause some humans to take an airplane flight to a far-flung place on the other side of the world so that they can take an 'eco-holiday' rather than a 'normal' holiday; the fact that that the holiday is labelled 'eco' helps to appease the guilt which arises from living a highly environmentally destructive lifestyle. In a similar vein, many humans feel good about themselves by taking their own reusable bags to the supermarket, where they then proceed to fill them to the brim with items that are covered with excessive packaging comprised of plastic and many other materials!! Then, many of these humans will load their shopping into their fossil-fuel burning car and drive home! Whilst the force to environmental sustainability is clearly at play in both of these situations, it is obviously utterly dwarfed by the force to environmental destruction. In the guilt-tinged age of the technological explosion, in the face of heightened ethical sensitivity, people often like to feel like they are 'doing something' to 'save the planet'. Of course, in reality, they are doing nothing of the sort! Saving life on Earth requires the human species to greatly increase its presence on the Earth, to modify more, and to power forward to the sufficient deployment of technology.

Both of these forces are inevitable components of human culture. The force to environmental destruction is currently overwhelmingly dominant and this is the way that things need to be at this stage of human cultural unfoldment. It is the overwhelming dominance of this force that results in the future bringing forth of the required sufficient deployment of technology. The force to environmental sustainability whilst still very weak will become progressively stronger until we reach the sufficient deployment of technology and the ensuing age of the 'spiritual explosion'. At this stage of Solar-Systic unfoldment everything changes because neither force is dominant. One of the hallmarks of the age of the 'spiritual explosion' is that the two forces that we have been considering will become perfectly balanced in human culture.

Technology & the emergence of the 'environmental crisis' – the four-stage process

We have briefly explored the fact that 'the environment' is the dominant theme in human culture in the year 2021. We need to ruminate on why this is so through an exploration of how technology is related to the emergence of the 'environmental crisis'. The term 'environmental crisis' has only very recently been coined and it has only been coined because of the existence of a great number of environmental problems. We need to ask: What is an environmental problem? An environmental problem is a modification to the Earth made by humans which has effects which human society considers to be unacceptable. If a modification is made to the Earth which has a non-human cause, then the effects of this modification do not constitute an environmental problem. And, if a human modification to the Earth has effects which human society considers acceptable, perhaps the planting of trees, for example, then this modification does not constitute an environmental problem. There is, of course, scope for disagreement concerning exactly which modifications made by humans are unacceptable; there might be an extremely widespread view in human society that a particular modification is acceptable, yet some humans might find this modification to be totally unacceptable. Nevertheless, there are clearly a plethora of diverse environmental problems. We can envision a sliding scale of environmental problems which has at one end those that are local and short-term, and which has at the other end those that are global and long-lasting. The term 'environmental crisis' has been coined because of the emergence, in the last century, of an increasing number of environmental problems that are towards the global and long-lasting end of the scale.

We can think of the two phenomena that we are concerned with here – technology and the 'environmental crisis' – as phenomena which emerged at a particular point in the journey that is the universal pursuit of ecstasy. There was a point in the Earth's history when it became technological; there was a point in the Earth's history when the 'environmental crisis' began. These events are linked because the former is the precursor of the latter – the emergence of technology results in a chain of events which lead to an 'environmental crisis'. There is a four-stage process of unfoldment. Firstly, there is the ecstatic bringing forth of technology. Secondly, technology takes on a momentum of its own and pervades the planet. Thirdly, technology becomes so pervasive that it causes modifications to the Earth that the master modifier would consider to constitute an 'environmental crisis' if it was aware of them. Fourthly, the human species comes to realise that the widespread deployment of technology has resulted in 'unacceptable' changes that constitute an 'environmental crisis'.

At the heart of this four-stage process is the fact that the human species only comes to realise that what it would consider to be an 'environmental crisis' exists, a long time after the unacceptable modifications that constitute the crisis exist. In other words, the bringing forth of technology entails an 'awareness of effects time-lag'. When a solar system brings forth technology, the master modifier is completely ignorant of the fact that such a bringing forth will also bring forth an 'environmental crisis'. Such ignorance is a necessity. After all, if the knowledge existed that the bringing forth and pervading of technology would inevitably bring forth an 'environmental crisis', then the human species would not allow technology to speedily and rampantly pervade! In this four-stage process of unfolding it is possible, and appropriate, to see humans as 'cosmic puppets'. The four-stage process is obviously intimately connected to the partial severance of ecstasy from its inner qualitative wisdom in the era of the 'technological explosion', which we explored earlier.

Given that the Earth bringing forth technology is an event of the highest magnificence, as it is the attainment of ecstasy, and given that the bringing forth of technology inevitably leads to an 'environmental crisis' (the four-stage process), we are left with the conclusion that the 'environmental crisis' needs to be seen in a positive light. If we were to imaginatively isolate the changes that constitute the 'environmental crisis' then they seem to be wholly negative. However, we are able to see the bigger picture. In reality, the modifications that humans have labelled the 'environmental crisis' need to be seen as deleterious side-effects of a much greater good – planetary life becoming technological and moving forward on the journey that is the universal pursuit of ecstasy. The 'environmental crisis' is a necessary stage of unfoldment as our Solar System progresses towards the sufficient deployment of technology and the age of the 'spiritual explosion'.

Two categories of environmental problems

We have explored how the 'environmental crisis' is constituted out of various environmental problems. One obvious question which raises itself, and which we need to address, is: *How can we resolve environmental problems?* There is a tendency for humans to come up with a 'blanket response' to this question. And this 'blanket response' is typically informed by an underpinning view concerning technology. So, on the one hand, there are those who have an underpinning view of technology as an 'evil', as a disruptor of pristine nature, which needs to be tamed and constrained, or even wiped off the face of the planet. This perspective leads to the 'blanket view' that the solution to environmental problems cannot possibly be more of what caused them; it has to be the opposite, the reigning in of technology. Those with this view will emphasise things such as recycling, reduction in resource use, efficiency savings, living a simple life and human population control, as the solution to environmental problems. On the other hand, there are those who have an underpinning optimistic view concerning the potential of technology to solve all kinds of problems and create a better future. This perspective leads to the 'blanket view' that technology can be the solution to environmental problems; for 'techno-optimists' the solution to all environmental problems is more of what caused the problem in the first place.

Our objective is to acknowledge that these two 'blanket views' exist and to clearly see why they are both wrong. When we look at the 'environmental crisis' from the perspective of the journey that is the universal pursuit of ecstasy, then we can clearly see that there are two categories of environmental problems. The first category contains just one environmental problem – human-induced global warming. The second category contains all of the other environmental problems. Human-induced global warming is in a category of its own because, as we have been exploring, it has an essential role to play in the journey that is the universal pursuit of ecstasy. It is a partial reversal in life on Earth's first line of defence, and this reversal plays a pivotal role in saving life on Earth through aiding the bringing into being of the sufficient deployment of technology. In other words, this partial reversal is required to bring forth life on Earth's main line of defence. All other environmental problems are not central to the universal pursuit of ecstasy; indeed, they can be seen as things which in themselves are wholly deleterious; they are effectively unwelcome side-effects of the striving for the sufficient deployment of technology.

What are the practical implications of the existence of these two categories? The primary aim of this book is to reveal why the only possible solution to the environmental problem in the first category – human-induced global warming – is a technological solution. This is because it is a partial reversal in life on Earth's first line of defence as it brings forth its main line of defence. This means that the force for human-induced global warming is not an isolated force. It fuses with the force for non-human-induced global warming to create a combined human/non-human force for global warming. As this force has both human and non-human causes we are not here in the realm of a traditional environmental problem; for, as you know, a traditional environmental problem is solely caused by human activities. A useful exploration of the fusion of these two forces can be found in this section: *Two Forces for Global Warming*. The partial reversal of the first line of defence is a trigger event which brings into being the mother of all perfect storms. The impending peril that life on Earth was facing from the force for non-human-induced global warming is ushered forward from the near future to the present due to the partial reversal. A fruitful elucidation of this trigger event can be found in this section: *The Oceanic Travelling Carbon Dioxide Gushes Out into a Vastly Changed World*.

In contrast to the first category, all of the second category environmental problems could have a technological solution, a non-technological solution, or a solution that is a mix of the two. In this category of problems there is scope for discussion, for flexibility and for diverse approaches to solve particular problems. It is the fact that the first category environmental problem only has a technological solution, whilst the environmental problems in the second category have many possible solutions, which generates the existence of these two categories.

The two non-cyclical one-way forces

In order to fully appreciate both that life on Earth is striving for the sufficient deployment of technology through the universal pursuit of ecstasy, and that there is an imminent need for this deployment, we need to come to appreciate how the two types of global warming interact on a thriving life-bearing planet. Let us recall what these two types of global warming are. Firstly, there is the global warming of our entire Solar System, which is driven by the increasing solar radiation that is emitted by our ageing Sun. This 'non-human-induced global warming' has become the enemy of life on Earth; life has for a very long time been combating it through its first line of defence (removing carbon dioxide from the Earth's atmosphere and storing it under the Earth's surface), but this line of defence is now in terminal decline and is approaching its expiration. Secondly, there is the global warming brought about by ecstasy as it initiates a partial reversal of the first line of defence, as part of the process of bringing forth the main line of defence (the sufficient deployment of technology). This is the phenomenon of 'human-induced global warming'. This is a crucial point of realisation: *human-induced global warming always needs to be seen in the context of non-human-induced global warming*. Why is this? This is because human-induced global warming is a partial reversal in life on Earth's first line of defence against non-human-induced global warming. Moving forward we need to come to fully appreciate both that human-induced global warming plays a role in bringing forth the main line of defence, and that such a bringing forth is required if life on Earth is to continue to survive and thrive despite the collapse of its first line of defence against non-human-induced global warming.

These two types of global warming are non-cyclical one-way forces which both exert upwards pressure on the Earth's atmospheric temperature. We need to reflect on a crucial difference between these two forces. The force for non-human-induced global warming fully reveals itself in the present. Whereas, the force for human-induced global warming has two elements: firstly, factors which are currently exerting upwards pressure (thereby revealing themselves in the present); secondly, factors which will inevitably exert upwards pressure in the future due to past events running their course and producing yet-to-be-manifested effects. It is very important to clearly appreciate the difference between the actuality of global warming and the existence of forces for global warming. To clearly see the forces at work is much more important than measuring the present actuality. Well, it is not just much more important, it is astronomically more important!

Let us start by considering non-human-induced global warming. We can see that this has been an increasingly powerful force which has been generating increasing pressure for the Earth's atmospheric temperature to increase ever since the Earth was formed. The fact that life on Earth has offset this force through its first line of defence has meant that there has been a lack of increase in the Earth's atmospheric temperature, a lack of actual global warming, despite this ever increasing force for global warming. In short, this immensely powerful force for global warming on the Earth has not, so far, turned into the actuality of global warming on the Earth. We are considering long timescales here, not hundreds of years, not thousands of years, but a stable atmospheric temperature over 3000 million years, fluctuating within a very narrow range, despite the force for non-human-induced global warming becoming immensely more powerful throughout this period. What is crucial is to appreciate the existence of the force, and its increasing potency over time. For, the lack of actual global warming can easily lead one to fail to recognise the forever increasing potency of the force.

Let us turn to human-induced global warming. A very similar thing applies; for, immense forces for human-induced global warming have been unleashed, but despite the immensity of these forces there has, comparatively, been very little actual global warming. The GMST has only risen by approximately 1 degree Celsius since the late 19th century. This might seem to be significant, but it is a trivially insignificant amount of actual global warming when compared to the immensity of the force for human-induced global warming that has been generated over this period.

We need to seriously ruminate on this. On the one hand, we have two immense forces for global warming which have been unleashed on the Earth; yet, on the other hand, we have a comparative lack of actual global warming. We need to fully appreciate why this is so. We need to come to really know these two forces for the warming of the Earth's atmosphere; we need to see all of their diverse aspects and manifestations. We need to start to comprehend the immensity of the forces that are currently exerting upwards pressure on the temperature of the Earth's atmosphere. We also need to appreciate that the actions of ecstasy in the past have unleashed chains of events which will exert upwards pressure on the temperature of the Earth's atmosphere in the future because these events have yet-to-be-manifested effects. One needs to clearly see all of this if one is to appreciate the need for the imminent bringing forth of the sufficient deployment of technology!

Let us explore the force for human-induced global warming in a little more detail. Human activities over time have continuously produced a stronger and stronger global warming force. The rainforests of the planet have been progressively destroyed and replaced with agriculture and urban sprawl. The history of fossil fuel use is one of forever increasing transfer from underground storage to the surface of the Earth. The human population size has progressively increased and this population has used increasingly more resources. The outcome of all of this is that as time has passed human activities have generated a continuously increasing force for human-induced global warming. We are here talking about the overall impact of human activities on the planet since ecstasy came into existence. This overall impact is a non-cyclical one-way force for global warming.

It is important to appreciate that it is possible for a non-cyclical one-way force to cease, to become a non-force. To talk of human-induced global warming as a non-cyclical one-way force is not to make predictions about the future; it is simply to recognise the reality of the past and the present; it is to say that this force has existed throughout human history and that it still exists today. We also need to recognise that over the past 200 years this force has become increasingly powerful. Yet, things change. When we reach the sufficient deployment of technology this force will no longer exist; for, the overall impact of human activities on the Earth will then be producing no force for the warming of the atmosphere.

Let us now turn to the force for non-human-induced global warming. When we consider the history of the Earth, from its formation to the present, then the overwhelmingly dominant force exerting upwards pressure on the temperature of the Earth's atmosphere is non-human-induced global warming. The source of non-human-induced global warming is the increasing amount of solar radiation reaching the Earth from the Sun. Since life arose on the Earth the amount of solar radiation reaching the Earth from the Sun has increased by a massive 25 %; yet, due to the success of the first line of defence, over this entire period the temperature of the Earth's atmosphere has been stable, fluctuating within a narrow range. Obviously, without the first line of defence, the temperature of the Earth's atmosphere would have increased very significantly! Such an increase would have decimated life on Earth. There are, of course, short-term cyclical variations in the amount of solar radiation that reaches the Earth due to the sunspot cycle. These variations are superimposed onto a continuous long-term one-way upwards process, the forever increasing amount of solar radiation being propelled towards the Earth. This one-way process gives rise to an unrelenting force for non-human-induced global warming. This force can be offset but, as long as our Solar System exists, it will never cease.

The interaction of the two forces

Recall that in stage four of the four-stage process, ecstasy comes to realise that the force for human-induced global warming exists. We now need to ask: How should ecstasy respond to this realisation? In considering the appropriate response to this question it would be very easy to make the mistake of believing that the force for non-human-induced global warming is an irrelevance. This is an easy mistake to make because the force for non-human-induced global warming operates over an immensely long timescale and it is a very slow and a very gradual process; whereas, in contrast, the force for human-induced global warming is something that is comparatively exceptionally new, very short-term in nature and apparently urgent. In order to avoid making this mistake one needs to come to realise that despite the slowness and the gradualness of the force for non-human-induced global warming, that this force has recently become immensely more potent on the Earth. The reason that it has become immensely more potent is that it has weakened the Earth's first line of defence to such an extent that it is close to overwhelming it; the first line of defence is approaching the point of eternal breakdown. A force which is very slow and very gradual can still have effects, at particular moments in time, which are exceptionally large and rapid. A term that has been coined that is applicable to this kind of situation is the phenomenon of the 'tipping point'.

The way that things unfold in a solar system means that the wearing out of a thriving life-bearing planet's first line of defence temporally coincides with that life bringing into being its master modifier / ecstasy which brings forth its main line of defence. This means that at the time that the force for human-induced global warming comes into existence, the force for non-human-induced global warming is an immensely potent threat due to the frailty of the planet's first line of defence. So, the coming into existence of the force for human-induced global warming means that life on Earth is in very grave danger from the force for non-human-induced global warming. These two forces compound and they inflame each other. This combination of forces means that the overall situation that life on Earth currently faces is very very serious; we are currently living at a time of immense danger for life on Earth.

The reason that this danger is so grave is obvious. The force for non-human-induced global warming is at the zenith of its potency at the same time as the force for human-induced global warming comes into existence. To say that the force for non-human-induced global warming is at the zenith of its potency is to say that it is posing its maximum threat to the continued flourishing of life on Earth, due to it being close to totally overwhelming life on Earth's first line of defence. If the first line of defence completely breaks down before the main line of defence is brought forth, then life on Earth would be decimated, so the force for non-human-induced global warming would no longer be a threat to the continued flourishing of life on Earth; the war would have been lost! Whereas, if the main line of defence has come into being, then non-human-induced global warming would no longer pose a threat to the continued flourishing of life on Earth. This is why it is in the immediate future that the force for non-human-induced global warming is at the zenith of its potency, its maximum threat to the continued flourishing of life on Earth. When it comes to the force for human-induced global warming, the fact that this force is currently, in the year 2021, also approaching its maximum potency, is something that we need to explore in more detail. What this means is that we need to explore why this force will, in the very near future, become exceptionally dangerous due to the coming to fruition of immense yet-to-be-manifested effects.

The need for human-induced global warming

At this stage we need to explore a little further why it is that both the force for human-induced global warming, and the actuality of human-induced global warming, are required in order to bring forth the sufficient deployment of technology. Why couldn't ecstasy simply bring forth the sufficient deployment of technology to combat the immense force that is non-human-induced global warming, without itself creating a force for human-induced global warming? Such a scenario is not impossible; for, it is possible to envision a world in which this occurs. However, the important thing to realise is that things don't actually turn out this way in reality. The universal pursuit of ecstasy, the natural desire of all things to attain 'increasingly intense pleasantness', 'motivates' things – from 'atoms' to 'humans' – to act in a certain way, and these actions result in an increasing ability of things to modify their surroundings as our Solar System unfolds. The coming into existence of ecstasy, the master modifier, is the zenith of this process. When ecstasy comes into existence its natural instinct is to keep on modifying, exploring and transforming; there is barely any thought or reflection concerning this. At the moment that the master modifier comes into existence and starts to master modify it does not ask itself: Should I/we be doing this? The master modifier doesn't stop modifying and ask itself: Should I/we be doing this?

Of course, the time eventually comes when the master modifier does ask this question of itself. However, the reason that it eventually comes to ask itself this question is that it has modified its host planet to such a massive extent that it barely resembles that which came before. The Earth needs to be covered with *mega-cities/concrete/factories/roads/cars/barren industrial agricultural expanses/ecstasy* before the master modifier asks itself these questions: Was this immense scale of transformation a good idea? What are the consequences of what we have done?! Before ecstasy made such a massive transformation there were no questions to be asked. So, these questions weren't asked. In short, it is inevitable that ecstasy will transform its host planet to such an extent that the first line of defence becomes partially reversed, through the pulling out of the ground, and burning, of masses of the fossil fuels that were safely stored there, before it asks itself any questions concerning whether or not this is/was a good idea. This is just the way that things unfold on a thriving life-bearing planet. In other words, the partial reversal of the first line of defence, caused by the mass utilisation of fossil fuels, is a required side-effect of life on Earth becoming technological; it is part of the technological birthing process. In other words, the creation of a force for human-induced global warming, which gives rise to actual global warming, is part of the process of life on Earth bringing forth that which it needs to survive – the sufficient deployment of technology. This is the prime reason why there is a need for human-induced global warming.

Now, of course, when the master modifier first comes to ask itself questions concerning the extent of its modifications to its planet, these questions and the answers they bring forth are tinged with guilt. What have we done?! Have we messed up the Earth? Are there too many of us? Are we going to master modify our way to extinction? Ecstasy comes face-to-face with the way that it has changed its planet and it doesn't like what it sees. The human species starts to tell itself that it has done something that is wrong and that the environmental problems that it has created through its modifications and transformations need to be addressed through stopping what it has been doing. We were wrong to chop down so much of the rainforests! We were wrong to use billions of plastic carrier bags! We were wrong to release all those fossil fuels! What on earth were we thinking? Or, more precisely: Why on earth weren't we thinking?! Replant the trees! Stop using the bags!! Stop using the fossil fuels!!!

In the year 2021 we are living in the midst of these questions and the midst of this guilt. This growing sense of guilt, this extremely limited view concerning the human place on the Earth, explains why so many people believe that the appropriate response to the phenomenon of human-induced global warming is simply to stop using fossil fuels. The flawed assumption underpinning this belief is that human modifications of the Earth are unwelcome and should be minimised. This initial response of ecstasy to the realisation of the extent to which it has modified the Earth is perfectly understandable; more than this, it is inevitable.

The questions that the master modifier eventually comes to ask itself, and the associated guilt, are in themselves very important. For, the beliefs that 'we have messed things up', that 'we are in a really sticky situation', that 'our backs are against the wall', are factors which can help to motivate the human species to bring forth the sufficient deployment of technology. In other words, the phenomenon of human-induced global warming – the actuality of human-induced global warming in tandem with the recognition of the existence of the immense force for human-induced global warming – and the associated concerns about its actual and potential future impacts (the immense death and suffering caused by climate change), can lead us down the path to the sufficient deployment of technology. This is the second reason why there is a need for human-induced global warming. In this regard, the recognition of the magnitude of the 'pent-up force' for human-induced global warming is of paramount importance. That is to say, ecstasy needs to come face-to-face with the currently existing forces that will inevitably exert enormous upward pressure on the Earth's atmospheric temperature in the near future, due to its past activities having yet-to-be-manifested effects.

At the time that it brings forth the sufficient deployment of technology, the overwhelming majority of the human species lacks the 'cosmic awareness' to appreciate that its destiny, its purpose, was always to bring forth the sufficient deployment of technology. This deployment benefits life on Earth, enabling it to win its war against the Sun. This deployment thereby enables states of ecstasy and states of heightened excitement/exhilaration to continue to exist in our Solar System. At the time that it brings forth the sufficient deployment of technology, the overwhelming majority of the human species also lacks knowledge of the immediacy and the immense potency of the threat that life on Earth was facing due to the two forces for global warming – human-induced global warming and non-human-induced global warming – compounding and inflaming each other. Eventually the guilt that comes to pervade ecstasy fades away and the human species comes to realise that it has a cosmic purpose. In this way the human species comes to realise that the transformations that it has made to the Earth have been for the benefit of life on Earth. The human species comes to realise that it is the precious ecstatic saviour of life on Earth when it becomes endowed with 'the realisation of cosmic purpose' in the age of the 'spiritual explosion'.

The 'pent-up force' for human-induced global warming

We need to clearly see the forces that currently exist, the forces that have been unleashed which are going to exert pressure for human-induced global warming in the future. To talk of such forces is to talk of the existence of time-lags between past ecstatic activity and a future increase in the temperature of the Earth's atmosphere; these are actions which have yet-to-be-manifested effects. We are here talking about human actions that have already taken place, which have set in motion chains of events whose 'final' atmospheric effects have yet to become manifest. Of course, the existence of these time-lags means that these forces for global warming might never lead to the actuality of global warming, because they might be counterbalanced by an opposing force which comes into existence in the future. Obviously, if there is no such counterbalancing force in the future, then these forces for global warming will inevitably lead to actual global warming.

We are here only interested in forces that are already 'pent-up'; these are forces that have already been unleashed but which have yet to reveal their global warming impacts. We are not interested in forces which come into existence through human actions that take place in the future. It is worth noting that, in the year 2021, at every moment of every day, new forces for human-induced global warming are currently being created by ecstasy, and some of these will become 'pent-up'. However, this newly created 'pent-up' force is, effectively, an irrelevance. We are solely interested in the 'pent-up' force that currently exists. This force is extremely potent, and potentially lethal, and it is a phenomenon that only ever comes into existence once in the journey of a thriving life-bearing planet, as it goes through its technological birthing process. This 'pent-up' force already exists, so there can be no room for hoping that it doesn't come into existence. The 'pent-up force' cannot be wished away!

The 'pent-up force' is set to significantly raise the Earth's atmospheric temperature in the very near future. It will dramatically increase the amount of carbon dioxide in the atmosphere, thereby enhancing the 'greenhouse effect'. We have explored how ecstasy has brought into being a partial reversal in life on Earth's first line of defence. In other words, human activities have resulted in an increase in atmospheric greenhouse gas concentrations. It is important to realise that, so far, the extent of this reversal has been minimal. This minimal change in greenhouse gas concentrations / atmospheric temperature which has already occurred, needs to be contrasted with the 'pent-up force' for future change that we are exploring here. In stark contrast to the tiny amount of actual global warming that has occurred as a result of human activity, the 'pent-up force' for human-induced global warming that has been generated is enormous!

What is this 'pent-up force'? You know that ecstasy has released an enormous amount of fossil fuels from underground storage, where they were not a force for global warming. If all of this released carbon dioxide moved to a new home in the Earth's atmosphere, then there would be no 'pent-up force' for global warming; there would just be actual global warming due to a stronger 'greenhouse effect'. However, this didn't happen. A colossal amount of the released carbon dioxide was immediately sucked into the ocean, where it is biding its time, waiting to move into the Earth's atmosphere in the very near future. This is why there is a time-lag between ecstatic activity and end result which constitutes the 'pent-up force' for human-induced global warming.

The part of the ocean that a colossal amount of the carbon dioxide that ecstasy has released is temporarily stored in is called the thermohaline circulation. The thermohaline circulation involves dense cold water sinking at high latitudes and travelling very slowly through the ocean depths until it eventually reaches the northern Indian Ocean and the northern Pacific Ocean where it resurfaces. This is not a quick journey! After carbon dioxide sinks into the thermohaline circulation it takes from 100 years to 1000 years to emerge into the atmosphere. This means that almost all of the carbon dioxide that has become stored in the thermohaline circulation since the start of the Industrial Revolution has yet to emerge into the atmosphere! Oh no! Oh no!! Oh no!!! When this carbon dioxide starts to be released *en masse* it is set to turn from a 'pent-up force' for human-induced global warming to a bringer of a large increase in the Earth's atmospheric temperature.

By around the year 3000 – when all of the carbon that is currently stored in the thermohaline circulation has been released – the conditions of the Earth could very easily be not suitable for the existence of ecstasy or for any other highly excited/exhilarated form of life. Of course, as we have been exploring a 'pent-up force', this would be the case even if humans were to stop using fossil fuels today.

The dramatic return into a vastly changed world

This 'pent-up force' is made immensely more dangerous, extremely more potent, because of the fact that the world that the travelling oceanic carbon dioxide will be gushing out into is a very different world to the world that it departed from. In this vastly different world, neither the land surfaces nor the oceans are such a welcoming home for atmospheric carbon dioxide. This means that when the long-travelling oceanic carbon dioxide gushes out of the thermohaline circulation it will accumulate in the Earth's atmosphere. Atmospheric carbon dioxide concentrations will spiral upwards. Why is the world to which the long-travelling oceanic carbon dioxide will be returning to so different? Why is it a vastly different world to the world that it departed from?

Let us start with the enormous transformation of the Earth's land surfaces. Since the start of the Industrial Revolution ecstasy has transformed these surfaces through mass deforestation. The extent and speed of the deforestation that has occurred in this period is mind-boggling. A ginormous amount of lush rainforest (which is a large and immensely valuable 'carbon sink') has been chopped down by chainsaws, and has been replaced by urban sprawl, concrete, industrial agriculture, mining operations and roads. Such a transformation makes it harder for life on Earth to keep its atmospheric temperature down.

Let us now turn to the transformation of the Earth's oceans. Since the start of the Industrial Revolution the Earth's atmospheric temperature has increased by approximately one degree Celsius. This warming of the atmosphere, caused by the ecstatic burning of fossil fuels, has caused the oceans to warm. This is of great importance because carbon dioxide is less soluble in warmer water. This warming of the atmosphere has also caused the melting of icebergs and glaciers in the North Atlantic and Greenland. This is of great importance because as the continuously melting freshwater pours into the Atlantic Ocean it slows down the thermohaline circulation. And a slower thermohaline circulation means less carbon dioxide in the oceans and more carbon dioxide in the atmosphere.

So, at the start of the Industrial Revolution, when the long-travelling oceanic carbon dioxide that we are exploring – that which emanated from the mass burning of fossil fuels – started to enter the thermohaline circulation, the Earth's land surfaces were home to immense lush rainforests, and the thermohaline circulation was vigorous. The rainforests were an immense 'carbon sink', and the thermohaline circulation was vigorously sucking in enormous amounts of carbon dioxide. One event – the Industrial Revolution – brought this world to an end and ushered in a vastly different world. This one event led to a transformation of the Earth's land surfaces through mass deforestation. This one event also simultaneously caused an enormous amount of carbon dioxide to build up in the thermohaline circulation, whilst also initiating a chain of events which are causing the thermohaline circulation to increasingly slow down! This one event also caused the oceans to be a less conducive home for carbon dioxide through raising the oceanic temperature. All of this, the coming into existence of this vastly changed world, means that when the long-travelling oceanic carbon dioxide starts to gush out of the oceans it will start accumulating in the Earth's atmosphere. Both the land surfaces, and the oceans, have been transformed since the long-travelling oceanic carbon dioxide departed. The mother of all perfect storms is about to hit the Earth in the coming decades and centuries.

We can clearly see that atmospheric carbon dioxide concentrations are set to spiral upwards for centuries. The Earth's atmospheric temperature is set to shoot upwards. Tipping points are set to be passed. Large-scale discontinuities, such as a runaway greenhouse effect resulting from the destabilisation of methane clathrate reservoirs, are set to be triggered. We are on the verge of the Earth becoming an inhospitable and barren place.

The next 300 years & the futility of cutting greenhouse gas emissions

Given that the amount of fossil fuels burned by humans has been enormously high in the past 100 years, and was also extremely high in the preceding 100 years, and given that roughly half of the released carbon dioxide was sucked into the thermohaline, from where it takes between 100 years to 1000 years to emerge into the atmosphere of the vastly changed world, you can quickly do the maths and work out that the next few hundred years are going to be interesting! The next 300 years are the time of real danger for life on Earth. This is the time when the force for human-induced global warming is at the zenith of its potency; this is mainly due to the coming to fruition of the 'pent-up force' that we have been exploring. However, in this period the 'non-pent-up' force for human-induced global warming also reaches its peak strength; this force peaks at its maximum potency, and then weakens due to the combination of the depletion of easily extractable fossil fuel reserves and the growing strength of the force to environmental sustainability. And, of course, the force for non-human-induced global warming is already approaching the zenith of its potency, having almost overwhelmed life on Earth's first line of defence. The combination of these forces, each of which is at the height of its potency, is something that the human species should be terrified about.

How should the human species respond to this potent danger that planetary life faces in the very near future? One thing is clear. This is the fact that reducing greenhouse gas emissions, reducing fossil fuel use, will not do anything to resolve the situation; this is because the source of the immense danger is a combination of non-human-induced global warming and the coming to fruition of the 'pent-up force' for human-induced global warming. In itself, such a reduction would seem to be a very marginal good thing, which could possibly buy us a tiny bit more time to bring into existence the sufficient deployment of technology. However, there is an immense danger that such reduction efforts might actually be catastrophically harmful. If ecstasy were to become obsessed with such reduction efforts, due to the deluded belief that such efforts herald the possibility of 'saving the planet', then life on Earth would be doomed! If ecstasy were to believe that such reductions are a solution to the problem that life on Earth faces in the next 300 years, then it would be gravely mistaken. This is the potentially deadly delusion that currently pervades ecstasy. If greenhouse gas emission reduction efforts take away resources and mental focus from the actual solution, the only solution, to the situation currently facing life on Earth – *active technological regulation of the Earth's atmospheric temperature which culminates in the sufficient deployment of technology* – then these efforts will lead to the obliteration of life on Earth.

Yet, life on Earth is not doomed! For, the universal pursuit of ecstasy will ensure that ecstasy brings forth the sufficient deployment of technology when it is needed. After all, we are little more than 'cosmic puppets'! The human species will come to see that due to the 'pent-up force' for human-induced global warming, alongside the ongoing futility of the attempt to slash fossil fuel emissions, that the active technological regulation of the Earth's atmospheric temperature is required in the very near future. Ecstasy will therefore do what it needs to do; it will do what it was brought into existence to do; it will do what it has been given the skills/abilities/talents to do; it will successfully bring into existence the sufficient deployment of technology. The human species will have then fulfilled its cosmic purpose; yet, it won't even know that it had such a purpose! At least, it won't know until our Solar System has unfolded to the point that is 'the day of the realisation of cosmic purpose'.

The technologies that are deployed & sufficiently deployed

It is time for us to explore the particular technologies that ecstasy will be deploying over the next few hundred years. It is important for us to appreciate that if there was only one type of global warming – human-induced global warming – then simply reversing what ecstasy has done, through deploying technologies to draw back down the carbon dioxide that has been released from underground storage, would be sufficient. In other words, ecstasy could perfect the technique of technologically pulling carbon dioxide out of the atmosphere and either using it (in concrete, for example) or storing it safely underground, as quickly as possible, and then it would be adequately prepared to respond when the carbon dioxide starts to gush out from the thermohaline circulation *en masse*. As long as this was successfully achieved, with the gushing carbon dioxide being sucked out of the atmosphere and used/stored, then the phenomenon of human-induced global warming would have been successfully dealt with. So, the force for human-induced global warming is something that can be counterbalanced by tinkering with the 'greenhouse effect'.

You have come to know that reality is very different to the woefully simplistic scenario that we have just considered. In reality, the situation that we face involves two different forces for global warming compounding and inflaming each other as they reach the zenith of their potency at the same time. We need to be thinking bigger. The technological devices that we have just considered are precious and valuable, yet they are limited in their effectiveness. We need to be making a full-blooded attempt to take control of the Earth's atmospheric temperature. Such an attempt, such control, is life on Earth's best hope of still being a thriving entity in the year 3000. We need to move into the realm of solar-deflection technologies.

'Carbon-sucking and using/storing' technologies are extremely limited in their effectiveness because they are simply a technological replacement for life on Earth's first line of defence. These devices mimic what non-technological life on Earth has been achieving for millions of years through its first line of defence. The limitations of these devices are thus patently clear. Our Solar System has reached the age where the Earth's first line of defence has passed its 'sell by date'. Life on Earth's first line of defence is fatally impaled. A thriving life-bearing planet requires a certain level of carbon dioxide to be in its atmosphere and we have reached the stage of Solar-Systic unfoldment where there isn't enough carbon dioxide left in the Earth's atmosphere to enable the smooth functioning of the first line of defence. This was the case before ecstasy initiated a partial reversal of the first line of defence. This means that whilst there are currently slightly higher atmospheric carbon dioxide concentrations due to the partial reversal, which can be taken out of the atmosphere by human 'carbon-sucking and using/storing' technologies, such a taking cannot be the ultimate solution to the situation that we face. Simply reversing the partial reversal, by tinkering with the 'greenhouse effect', will, at best, just buy us a tiny bit more time. If we are to enable life on Earth to survive and thrive we need to be making a bigger and a bolder response. We need to embrace our cosmic purpose! We need to embrace our destiny!

The 'carbon-sucking and using/storing' technologies that we have just been considering will be widely deployed, and they will fulfil their purpose of helping to bring forth the sufficient deployment of technology. These technologies are obviously not powerful enough to be the sufficient deployment of technology. These technologies can actively control the temperature of the Earth's atmosphere, but they only do this in a limited and short-term way, through providing a temporary 'patching up' of the fatally impaled first line of defence. It is because these technologies only tinker with the 'greenhouse effect' that they are extremely limited in their effectiveness. A technological 'patching up' of the first line of defence, effectively only reverses the changes that ecstasy has made in order to enable the bringing forth of the main line of defence; it does nothing to address the underlying problem that we face, the problem that generated the need for the main line of defence in the first place. It matters not a jot whether or not the first line of defence is non-technological or is bolstered by technology, the first line of defence will soon be incapable of protecting life on Earth from the force for non-human-induced global warming.

Let us recap and draw some conclusions. The breakdown of life on Earth's first line of defence temporally coincides with the coming into existence of ecstasy and its technological fruits. Whilst, the bringing into existence of technology involves the creation of a partial reversal of the first line of defence, a reversal which is itself reversed by the technologically bolstered first line of defence ('carbon-sucking and using/storing' technologies). In other words, the enormous decline in atmospheric greenhouse gas concentrations which is the hallmark of the first line of defence turns into a small increase, which is followed by an offsetting decrease, as technology is brought forth on a planet. The bringing into existence of technology also creates a massive force for human-induced global warming, which is dominated by the 'pent-up force' that is created by the temporary storage of an enormous amount of carbon dioxide in the thermohaline circulation. This means that the force for human-induced global warming inevitably reaches the zenith of its potency at the same time as the force for non-human-induced global warming is at the zenith of its potency (its maximum threat to the continued flourishing of life on Earth). We need to see the threat that life on Earth faces as a combined threat, rather than as two different threats from two separate forces for global warming. It is this combined threat which needs to be focused on and addressed. This combined threat can only be dealt with through the sufficient deployment of technology.

The threat that life on Earth faces over the next 300 years is an overall threat, a combined threat, from the two forces for global warming being at their maximum potency at the same time. If this overall threat is not soon counterbalanced by technologies that are deployed and sufficiently deployed, then it will decimate life on Earth. In other words, we are approaching the time of the unfolding of life on Earth which is 'make or break' time. Either life on Earth will be overwhelmed by the combined force for global warming and be decimated, or it will pull through and enjoy a long and glorious future thanks to the sufficient deployment of technology. Whilst technologies that tinker with the 'greenhouse effect' are magnificent, they aren't sufficient. These technologies are valuable because they have the potential to buy ecstasy a tiny bit more time to bring forth the sufficient deployment of technology. These technologies are also valuable because they are a 'stepping stone', in terms of both concrete physical preparation and conceptual acceptance, to the sufficient deployment of technology.

The threat that we face over the next 300 years comes from a delicate interplay between forces in our unfolding Solar System, an interplay that results in the two different forces for global warming temporally combining so as to present an exceptionally potent threat to life on Earth. Creating part of this potency, and then offsetting the potency in its totality, in order to keep the Earth a habitable place for the wonderful life-forms that exist in our Solar System, is the purpose for which ecstasy was brought forth into existence. Don't reach the wrong conclusion. Don't believe that the danger that life on Earth faces comes solely from human activity and that it can therefore be combatted through technologies that tinker with the 'greenhouse effect'. Offsetting the exceptionally potent threat that life on Earth currently faces requires technologies that block incoming solar radiation, thereby preventing it from reaching the Earth's atmosphere. These technological devices are the sufficient deployment of technology.

Such devices are the only way to deal with non-human-induced global warming. Such devices might be the only way to effectively deal with the combined threat emanating from the two forces for global warming; the combined threat that is about to hit the Earth. This is because when the 'yet-to-be-manifested' effects come to fruition, when the temporarily stored carbon dioxide starts gushing out of the thermohaline circulation and into a vastly changed world causing an escalating 'greenhouse effect', incoming solar radiation can then be deflected in a controlled, effortless and speedy manner so as to keep the Earth's atmospheric temperature stable. If we were to put our trust in less powerful technologies, then we would be playing with fire; we would be putting the life that has arisen on the Earth at risk of obliteration due to our laziness and lack of ambition. Thankfully, we are not lacking in drive, energy and ambition. We are ecstasy. We are the master modifier of our Solar System! We are the saviours or life on Earth!!

Worrying about the wrong things

We have just considered two different types of technological deployment – 'carbon-sucking and using/storing' technologies and the sufficient deployment of technology. There are a great many humans alive today that are worried, even terrified, about the prospect of the deployment of these technologies on the Earth. It is worth exploring why these humans are worrying about the wrong things.

Let us first consider the deployment of 'carbon-sucking and using/storing' technologies. We have obviously already used technological devices to move carbon from underground storage to the Earth's oceans and atmosphere. Now we can use technological devices to move carbon in the reverse direction – from the Earth's surface back to underground storage. Such a basic form of atmospheric temperature regulation, simply reversing what we have previously done, is not something that one should be worried about. This reversal leads to stabilisation, not destabilisation. It is the prospect of a lack of technological control of the atmospheric temperature that one should be deeply concerned about. One should be worried by the fact that ecstasy has used technological devices to move a massive amount of fossil fuels from underground storage to the surface of the Earth, and in so doing has massively destabilised the Earth's biogeochemical cycles. One should most definitely be very worried by the fact that the effects of this destabilisation have yet to become manifest; this is because an enormous amount of the released carbon dioxide has entered a temporary home in the thermohaline circulation, from where it is biding its time waiting to gush forth into a vastly changed world. One could very reasonably be even more worried by the fact that the vast majority of humans are not aware of the extreme level of the destabilisation that currently exists. The last thing that one should be worrying about is the deployment of 'carbon-sucking and using/storing' technologies to reverse what ecstasy has previously done. Why on earth should one worry about the meticulously planned reversal of a destabilising process which was unplanned and chaotic! 'Carbon-sucking and using/storing' technologies can help to bring stabilisation to that which is currently massively destabilised. These technologies are not powerful enough to be a solution to the situation that we face; for, they are very limited in their effectiveness. Yet, they do have the potential to provide a limited short-term stabilisation. Such technologies are certainly not dangerous. They are not something to be worried about! Come on! If you are going to worry, at least worry about the right things!

It is time for us to consider the sufficient deployment of technology – technology that blocks incoming solar radiation, preventing it from reaching the Earth's atmosphere. Why might one be worried, even terrified, about the prospect of this technological deployment? Such a deployment undoubtedly entails the possibility of having some unforeseen effects, effects which could have significant impacts on the climate in particular parts of the Earth. Such effects could result in disputes and conflicts between countries, which would be unfortunate. However, we are clearly focused on the bigger picture, the interests of planetary life rather than the interests of particular instantiations of life. If there was no need for the sufficient deployment of technology, then it would be better to not deploy this technology and thereby to avoid the possibility of undesirable unforeseen effects. However, if you are worried about such possible effects to such an extent that you think that the sufficient deployment of technology should not be deployed, then you are missing the bigger picture: without the sufficient deployment of technology life on Earth would be decimated. Also, it is worth keeping in mind that following the sufficient deployment of technology we will be speeding into the age of the 'spiritual explosion'. This is important because in this epoch any undesirable unforeseen effects which affect particular areas/countries will be increasingly dealt with in a cooperative, compassionate and communal manner.

What you should really be worried about is what would happen to life on Earth in the absence of the sufficient deployment of technology. Decimation would await! Annihilation! No more states of heightened excitation/exhilaration would exist in our Solar System. No more ecstasy. Solar-Systic doom! The Earth would soon be home to nothing more than extremely simple life-forms which were unable to thrive, unable to complexify along the path towards ecstasy, because the atmospheric temperature of the Earth would be too hot for life to thrive. The future would be no different; for, it would entail only forever increasingly higher atmospheric temperatures due to the continued global warming of our Solar System. The Earth would still be home to simple life-forms, but they would effectively be living out their final days. The journey that is the universal pursuit of ecstasy would be forever thwarted in our Solar System. What a shame! Why don't you start to worry about this possible future for our planet? Please, don't worry about the sufficient deployment of technology! It is this deployment which is the glorious saviour of the wonderful highly developed life that currently resides on the Earth.

The stories that we tell ourselves

It will be fruitful for us to reflect on the fact that we like to tell ourselves stories concerning our significance or insignificance in the big scheme of things. On the one hand, we can tell ourselves that our species is totally insignificant as far as life on Earth is concerned, as far as our Solar System is concerned, and as far as the Universe is concerned (we are 'just one species among many' in a purposeless Universe). On the other hand, we can tell ourselves that our species is of immense positive significance to life on Earth, our Solar System and the Universe; the most important species that has ever existed and will ever exist. It is usual for a multitude of variants of these two types of stories to coexist, whilst one particular story is likely to be dominant at any particular moment of cultural unfoldment. Furthermore, most of these stories will contain at least a grain of truth.

Some of these stories are tied up with God, the stories presenting a view that we as a species were created 'in the image of God' and that we are raised up above all of the other life-forms of the Earth in terms of cosmic significance. These stories include the Biblical account of Adam and Eve and the view that humans have rightful dominion over all of the other life-forms that inhabit the Earth. They also include the Biblical tale of Noah's Ark which involves the human species using its ability as a highly skilled modifier of its surroundings to create an Ark which saves non-human Earthly animals from a great flood, a flood which would have otherwise caused their extinction. These are stories of positive human cosmic significance which entail that the human species is of immense value and importance in the big scheme of things.

From the perspective of the universal pursuit of ecstasy we can see the strands of truth in these stories of human cosmic significance. We can see that the *human species = ecstasy* is the most important life-form on the Earth, that the interests of life on Earth have been served by human domination/dominion over that which is non-ecstatic, and that the ability to increasingly modify, which reaches its zenith in ecstasy, is of immense importance. We can, of course, see the strands of truth in tales such as Noah's Ark. From our perspective, we can see that the essence of the Noah's Ark tale is that the human species is the saviour of the non-human life-forms of the Earth because of its master modification abilities. However, rather than saving these life-forms from a 'great flood', the reality is that they are being saved from an increasing atmospheric temperature that is caused by non-human-induced global warming. The 'Ark' that saves these life-forms is thus not a boat; it is the sufficient deployment of technology. Having said this, there is actually a very real sense in which the sufficient deployment of technology could save the life-forms of the Earth from a 'great flood', because this deployment can prevent a large rise in sea levels caused by the warming of the Earth's atmosphere. A large rise in sea levels would flood great swathes of the Earth's land surface areas and in so doing it would wipe out the life-forms that live there. So, in the absence of the sufficient deployment of technology, it would only be a matter of time before a plethora of life-forms on the Earth would actually be decimated by a 'great flood'!

Currently, in the year 2021, the dominant story that pervades human culture is very different from these stories of positive human cosmic significance. The dominant story of our time results from the fact that we are living in the midst of the 'technological explosion'. As ecstasy in the age of the 'technological explosion' we inevitably see ourselves as separate from that which is non-ecstatic, we see ourselves as fundamentally alien to the rest of the Earth. We don't call the non-human life-forms of the Earth our 'brothers/sisters', we call these non-ecstatic life-forms 'animals' and this is not a term of endearment! It is a term that is often used to degenerate and offend other humans, such as in the expression: *'You are nothing but a low-down dirty animal, a complete and utter waste of space!'*

The separateness that underpins the 'technological explosion' has given rise to stories that entail that the human species has no positive cosmic significance or importance. Most of these stories take a negative view of the human species through portraying – due to environmental destruction and the utilisation of non-human life-forms – the human species as the enemy of life on Earth, an unwelcome guest on a previously thriving planet. Other stories take a neutral view of the human species, through envisioning that various life-forms have inhabited the Earth – dolphins, dodos, chipmunks, humans, dinosaurs, mosquitoes, bonobos, ants, and so on – and that there is no significant difference or division between these life-forms. According to this way of thinking, there are just different life-forms living their lives and their activities are neither negative nor positive. In the midst of separateness and alienation one cannot see the bigger picture; one is blinded to cosmic purpose and significance, so one assumes that these things do not exist.

The 'technological explosion' inevitably followed the Scientific Revolution and these events played a role in transforming the previously dominating stories of human cosmic significance. These previously dominating stories were inspired by God, and in the age of technological mastery and of rapidly increasing scientific understanding of the way that the Universe operates, these stories have inevitably waned. As these God-inspired stories of human cosmic importance have waned, the vacuum has been filled by scientifically-inspired stories of the human species as 'just one species among many' which has been brought forth through 'natural selection'. If the human species is 'just one species among many' then the human species obviously has no cosmic significance as far as life on Earth, our Solar System, or the Universe, is concerned. Whilst these stories of cosmic insignificance are wholly false, their dominance in the age of the 'technological explosion' is an inevitable and essential part of the journey that is the universal pursuit of ecstasy.

Whilst there are a plethora of stories that we tell ourselves, there is only one truth – 'the way that things are'. As we explored earlier, the mystic can gain access to this truth in advance of its widespread realisation. Of course, as you are well aware, the widespread realisation of this truth, the widespread appreciation of the nature of human purpose, of human cosmic significance, of why the human species has acted the way that it has, of the existence of the universal pursuit of ecstasy, occurs when our Solar System has unfolded to the point that is 'the day of the realisation of cosmic purpose'.

A sense of specialness

The stories that we create concerning our cosmic importance or unimportance as a species operate in the realm of rationality. These stories arise from the need of the human species, as a highly rational being, to try and make sense of its own existence. In this segment our objective is to consider another aspect of ecstasy; we are leaving the realm of the rational and exploring the 'sense of specialness' that individual instantiations of ecstasy develop. *Every instantiation of ecstasy has a sense that the human species is special.* What exactly does this mean? This sense is an inner gut feeling, an inner knowledge, that the human species is not just another species of animal; a sense that there is a fundamental division of some kind which has the human species on one side and all of the other planetary life-forms on the other side. We need to ask: Where does this 'sense of specialness' come from? If one goes far enough back in the life of an individual human then one will arrive at the moment when this sense that the human species is special first arose; babies do not have this sense. When infants start to explore their surroundings they don't have a sense that the human species is special, that it is divided from, and raised up above, all of the other life-forms of the Earth.

The arising of this 'sense of specialness' occurs when two things have happened. Firstly, an infant needs to become ecstasy; it needs to start purposefully utilising technology. Secondly, this particular instantiation of ecstasy needs to reach what we can very loosely call a 'broad overview' of their surroundings. A 'broad overview' of one's surroundings entails a fairly wide range of knowledge about those surroundings; not detailed knowledge about any particular thing, just a little knowledge about lots of things. So, to put it extremely crudely, if one knows that there are countries and continents, mountains and oceans, fish and birds, insects and bacteria, planets and stars, boats and cars, airplanes and satellites, chimpanzees and dolphins, mobile phones and laptop computers, and cats and dogs, and one can loosely 'join the dots' between these things, then one has a 'broad overview' of one's surroundings. 'Joining the dots' means knowing that we live on a planet, a planet that has countries, oceans and mountains, a planet that is home to a plethora of life-forms, a planet that is surrounded by other planets and stars, and knowing that the human species has created things such as satellites, airplanes and mobile phones, whilst cats/dogs/chimpanzees/dolphins haven't done anything even remotely like this.

When these two things have happened an individual human will have a sense that the human species is special. This is because humans have glaring abilities that other things lack; abilities of a completely different scale of magnificence; abilities of a totally different magnitude to the abilities of anything else! If a human grows up surrounded by intricately moulded human-created 'things', such as designer clothes, a nice house, books, cars, airplanes, skyscrapers, kitchen appliances, televisions, mobile phones and computers, whilst being aware that the non-human life-forms that they share the Earth with have, by comparison, barely modified their surroundings, then it is hardly surprising that a sense that the human species is special arises within that human. This 'sense of specialness' is not something that is rationalised because, against the backdrop of one's 'broad overview' of one's surroundings, it is something that automatically arises from one's interactions with one's surroundings. In the age of the 'technological explosion' this 'sense of specialness' is largely dormant, meaning that it doesn't lead to rationalisations, to deep questioning, in the overwhelming majority of humans. When our Solar System unfolds to the point of 'the day of the realisation of cosmic purpose', then the 'sense of specialness' will cease to be dormant and will come to the fore; humans will collectively exhort: *'Of course we are exceptionally special as a species, the saviours of life on Earth; we already knew this!'* From the moment of this widespread exhortation, the 'sense of specialness' will be fully embedded in the realm of the rational.

Eagles & dolphins

Is it possible that non-ecstatic life-forms could have a sense that their 'species' is special? Let us consider an eagle. As the eagle flies majestically through the sky it looks down and sees humans who are rushing around in their nine to five working day, stuck in traffic jams, jostling their way through crowds, getting stressed, robbing each other, assaulting each other, arguing with each other and blowing each other up with explosive devices that they have created. The eagle also looks down and sees other life-forms that seem to be inferior, either because they are 'stuck' to the surface of the Earth or because they cannot fly as majestically as eagles can. It is easy to imagine that the eagle, who spends its day joyously gliding around in the sky, might have a sense that its 'species' must be special, the most important life-form on the planet. Eagles spend their time majestically soaring through the atmosphere, having a ball, whilst the 'human slaves' are working till they drop, getting stressed and living lifestyles which lead to widespread illness and disease. Let us consider a dolphin. As it communicates through echolocation, as it jumps out of the ocean and spins around in the air, as it saves a human from near-certain death, it is easy to imagine that it might have a sense that its 'species' must be special, the most important life-form on the planet.

What conclusions can we draw from such considerations? If we were to accept that a non-ecstatic life-form could have a sense that its species is special, then it would mean that having a 'sense of specialness' and actually being special are two very different things. For, there is no fundamental distinction of cosmic significance which has eagles on one side and all non-eagle life-forms on the other side. There is also no fundamental distinction of cosmic significance which has dolphins on one side and all non-dolphin life-forms on the other side. The only fundamental distinction of cosmic significance within life on Earth is between ecstasy and non-ecstatic life-forms. Whilst it is good to explore various possibilities, in reality it is only ecstasy that has a 'sense of specialness', a sense that it's species is fundamentally different, fundamentally superior, to all of the other life-forms on the Earth. This means that in reality the 'sense of specialness' matches the actuality of specialness.

There is one more issue that we should briefly address. Non-ecstatic life-forms do not have a sense that their 'species' is special, but is it possible that these life-forms could have a sense that ecstasy is special? We shouldn't rule out the possibility that some non-ecstatic life-forms have such a sense! For, the cosmic significance of ecstasy, the immense value of ecstasy as saviour of life on Earth, the qualitative vibrancy of ecstasy's concrete existence as ecstatic things loom out of their surroundings when they are sensed by other things, is so crystal clear, so radiant, so irrefutably and blindingly sublime, that it can be sensed by some of the non-ecstatic life that needs ecstasy to save it.

The many relationships between ecstasy & non-ecstatic life-forms

It is time for us to ponder the various relationships that exist between ecstasy and non-ecstatic life-forms. Firstly, there is the temporal transitional relationship between ecstasy and the non-ecstatic life-forms which were the immediate precursors of ecstasy – the 'non-human biological humans' which transformed into ecstasy when they brought forth technology. We have already explored this relationship and there isn't anything else that needs to be said here. Secondly, there is the present day relationship between ecstasy and 'non-human biological humans'. So, babies aren't part of ecstasy, and adults that live as recluses and/or in extremely remote places, and who have never had anything to do with technology, are also not part of ecstasy. The third type of relationship is that between ecstasy and all of the other life-forms that inhabit the Earth. In one sense, ecstasy is fundamentally opposed to these other life-forms, because it is the master modifier, the zenith of the evolutionary progression of life on Earth, the saviour of life on Earth, that which sees itself as fundamentally opposed to the non-itself, that which sees itself as non-natural (ecstasy creates the word 'natural' to refer to everything that is either not it, or not created by it), and that which has a unique 'sense of species specialness'. In another sense, this opposition is not quite so fundamental; for, some aspects of the opposition are created within ecstasy itself. This means that part of this opposition can be seen through, it can be penetrated and revealed for what it is: a cosmic tool that enables ecstasy to fulfil its cosmic purpose for the benefit of life on Earth.

In reality, all ecstatic and non-ecstatic life-forms are brothers/sisters. Yet, at the same time, there is obviously a very stark division within the life-forms of the Earth; for, ecstasy is the most important part of life on Earth. This is because ecstasy, as the master modifier among apprentice modifiers, has the ability to bring forth the prize technological fruit that life on Earth desperately needs – the sufficient deployment of technology. It is the purpose of ecstasy to bring forth the sufficient deployment of technology and thereby enable the continued thriving of life in our Solar System, in the face of the threat from non-human-induced global warming. Given that ecstasy is the saviour of life on Earth, and given that the act of saving requires in-depth knowledge of the Earth and the mass mobilisation and transformation of the Earth's resources, we can clearly see the rightness of the domination of the ecstatic over the non-ecstatic. In other words, it is in the interests of non-ecstatic life to be utilised by ecstasy, because this ensures the continued existence of planetary life. Of course, non-ecstatic life might not realise that such utilisation is ultimately in its interests, ensuring its future survival, in a similar vein to the way in which ecstasy currently lacks a widespread appreciation of its cosmic purpose.

We are here focusing, as we need to, on planetary life / life on Earth. If we were to focus in on an individual non-human life-form, then we could make a good case that its being utilised by ecstasy is not in its interests. After all, how could life in a factory farm as a food source for ecstasy, or life in a vivisection laboratory, possibly be in the interests of a thing? However, we have escaped this individualistic view of things; we are starting to see things as they are, as collectives, as wholes within wholes, as differentially embedded interconnected diasporic entities that unfold through time. It is always about the interests of the whole; the interests of the parts are always subservient to the interests of the whole. In other words, individual life-forms serve the interests of life on Earth.

Thankfully, our Solar System unfolds, things develop and change, and what was for the best in the past is not necessarily for the best in the future. We are currently living at the time when the relationship between ecstasy and the non-human life-forms of the Earth is, ever so slightly, starting to change. The relationship is very slowly starting to move from 'rightful domination' to 'rightful respect'. When the age of the 'spiritual explosion' reaches its zenith, with the arrival of 'the day of the realisation of cosmic purpose', then ecstasy will have a very different relationship with the rest of the life-forms of the Earth. All life-forms will be valued and respected as equals, as brothers/sisters that are jointly striving to stay in existence.

There is one more thing to be said concerning the relationship between ecstasy and the rest of the life-forms of the Earth. This is the issue of the possession of attributes. The human species has, for a very long time, been mentally grasping around looking for an attribute which can make it 'superior' to the other life-forms of the Earth. Ecstasy has been driven to do such grasping in order to justify to itself its exploitation of these life-forms. So, ecstasy has told itself that non-human life-forms are mere machines. What an absurdity! Such a view, whilst absurd, actually has its uses because it helps to propel life on Earth forward on its journey along the universal pursuit of ecstasy towards the sufficient deployment of technology. Ecstasy has also told itself that non-human life-forms, the 'mere animals', lack awareness, lack language, lack rationality, lack culture, lack the ability to use tools, lack emotions, lack creativity, lack the capacity to suffer, lack 'consciousness', lack..., lack..., lack..., the list goes on and on! We can clearly see the absurdity of all of this, whilst also appreciating its necessity in the age of the 'technological explosion'. In reality, a plethora of diverse non-human Earthly life-forms have all of the attributes listed above. The only thing that meaningfully distinguishes the human species from the rest of life on Earth is the fact that the human species is ecstasy, the master modifier of its solar system.

The journey that is the universal pursuit of ecstasy

Let us remind ourselves what it means to talk of the journey that is the universal pursuit of ecstasy. This journey involves the entire 'hierarchy of spatio-temporal embeddedness' striving for 'increasingly intense pleasantness', which equates to an enhanced modification ability, and which has as its destination the bringing into being of *ecstasy = the master modifier*.

When ecstasy has been brought into being the journey that is the universal pursuit of ecstasy does not end; it continues as technology gradually pervades until the age of the 'technological explosion' comes into being, thereby bringing forth increasingly intense ecstatic states. The age of the 'technological explosion' culminates with the bringing into being of the ecstatic prize fruit that is the sufficient deployment of technology. To bring forth this deployment is the purpose of the human species. The journey continues with the coming into being of the age of the 'spiritual explosion'. The hallmark of this age is the widespread maximisation of individual human potential and fulfilment through individuals following the path of spiritual development. This age culminates with 'the day of the realisation of cosmic purpose'.

In one sense, 'the day of the realisation of cosmic purpose' is the end of the journey that is the universal pursuit of ecstasy. In another sense, the journey is never-ending. This is because all things, both ecstatic things and non-ecstatic things, are perennially striving for 'increasingly intense pleasantness'. There is a never-ending movement towards higher states of excitement/exhilaration, greater levels of fulfilment, and the consequent coming into being of a Universe that more closely approximates both L* and E*.

Darwin's theory of natural selection

Ecstasy is the zenith of the evolutionary progression of life on Earth. Given that this is so, we should explore how this relates to Charles Darwin's theory of natural selection. The theory of natural selection is not wrong; it is just extremely limited in its scope. Let us imagine that a complete account of why the evolution of life on Earth has followed the trajectory that it has is represented by an iceberg. Charles Darwin's theory of natural selection would be part of this account, part of this iceberg. However, it would only apply to the tip of the iceberg, the tiny portion of the iceberg which is visible above the surface of the water; the overwhelming mass of the iceberg, which lies below the surface, is still to be accounted for. The fundamental drivers of evolution are outside the scope of Darwin's theory.

We can also think of natural selection in terms of a tree. Imagine that the evolution of life on Earth is a tree which grows from a miniscule size to a massive height. Natural selection applies to the branches and leaves, but not to the heart of the tree – the massive trunk. As the trunk of the tree moves upwards towards the sky it provides a direction from which the branches can spread out. This trunk, which equates to the evolutionary trajectory of life on Earth, is forged/propelled by the strivings of the qualitative states that comprise the universal pursuit of ecstasy, as they strive for 'increasingly intense pleasantness', in tandem with the cycles of qualitative unfoldment of our Solar System which are associated with the swirling of the planets around the warming Sun. We earlier explored this tandem process, which involves continuous striving within the context of cycles of qualitative unfoldment.

The theory of natural selection needs to be viewed from within the perspective of the universal pursuit of ecstasy. Considering the theory of natural selection in isolation is like focusing solely on the icing in blissful ignorance of the cake. The universal pursuit of ecstasy gives an explanation of why things, both non-living things and living things, come together and form new things. The transformation of things, the bringing forth of new things, is driven by states of qualitative feeling seeking 'increasingly intense pleasantness'. Things binding together in certain ways, life increasing its complexity, its ability to modify and control its surroundings, entails 'increasingly intense pleasantness'; that is, it entails states of heightened excitement/exhilaration. This process of coming together, of transformation, is constrained by the stage of maturation of Solar-Systic unfoldment that exists at a particular moment in time.

We need to appreciate that subtleness of the cycles of unfoldment in our ageing Solar System, as the planets swirl around the Sun. Firstly, there are cycles which seemingly get repeated over and over again. For example, the Earth taking 365 days to revolve around the Sun, and particular planets coming into and out of a series of various alignments with each other. As we have explored, particular stages of these repetitive cycles are associated with a particular type of qualitative state dominating great swathes of our Solar System. Secondly, it needs to be appreciated that these cycles of unfoldment are never ever truly repeated. Each cycle of unfoldment is subtly different from the preceding one; each cycle of unfoldment represents a gradual maturation and unfoldment of our Solar System. As our Solar System matures, greater opportunities for the emergence of heightened states of excitement/exhilaration, and ultimately for ecstasy itself, are brought into being. In other words, as the cycles of Solar-Systic unfoldment keep 'repeating', the conditions come into being which enable the journey that is the universal pursuit of ecstasy to continue; the strivings of all that is are gradually enabled to be increasingly successful.

The process that we have just explored, the strivings of the qualitative states that comprise the universal pursuit of ecstasy, which are gradually enabled and influenced by the cycles of Solar-Systic unfoldment, is the mechanism which has caused life on Earth to take the trajectory that it has from simple beginnings to the attainment of ecstasy. This mechanism is the trunk of the tree of both Solar-Systic evolution and the evolution of life on Earth; it is also the branches of the tree and the leaves of the tree. The trunk of the tree represents the bringing of things together and the forging of new types of life; that is to say, it represents the bringing forth of new movement patterns, new forms of modification ability, heightened states of excitement/exhilaration. We can see that natural selection is only a very minor part of the universal pursuit of ecstasy; it only applies to the branches and leaves of the tree, to the realm of small variations of that which has already been forged through the universal pursuit of ecstasy. All of this will become widely appreciated when our Solar System has unfolded to the point that is 'the day of the realisation of cosmic purpose'.

Dinosaurs, asteroid strikes & the singular path to ecstasy

It will be useful for us to delve deeper into the evolutionary trajectories of life on Earth. With this objective in mind, it will be fruitful for us to explore the topic of the extinction of the dinosaurs from the perspective of the universal pursuit of ecstasy. There is a widely held view that the dinosaurs were wiped out by the collision of an asteroid with the Earth, and that if this asteroid had just missed the Earth that the way that life on Earth unfolded would have been very different. According to this view, if the asteroid had missed the Earth the dinosaurs could still be 'ruling the planet' and ecstasy might never have evolved. This view is wrong. Even if the asteroid in question had missed the Earth the dinosaurs would still have gone extinct and ecstasy would still have evolved. But could the asteroid in question actually have missed the Earth?

We can imagine two scenarios, the first of which involves the asteroid in question hitting the Earth, the second of which involves this asteroid missing the Earth. Which scenario comes to pass does affect the evolution of planetary life, but this impact is very limited in the big scheme of things. This is because the direction of the unfolding of life on Earth is primarily driven by the striving of life itself, the striving that is the universal pursuit of ecstasy in tandem with the cycles of Solar-Systic unfoldment. Events such as small and medium-sized asteroid strikes simply modify this process at the edges; they do not fundamentally change the trajectory of the unfolding of life.

To believe that the cosmic bringing forth of ecstasy was inevitable is to believe both that the evolutionary trajectory of life on Earth was always heading towards ecstasy, and that this pursuit would indubitably be successful. There are seemingly a plethora of possible paths to this destination. One can imagine life on Earth striving for ecstasy and its route to this destination being modified by asteroid strikes and by other events such as massive volcanic eruptions. One can also imagine that the striving for ecstasy might have ended in failure, due to the collision of a massive asteroid with the Earth causing the obliteration of life on Earth. However, whilst this is a fruitful thing to envision, the striving for ecstasy actually occurs within an unfolding Universe that is predetermined in the nature of its unfolding.

Solar-Systic unfoldment up to the point at which rationality/awareness was brought forth couldn't have been any different; in this stage of the journey that is the universal pursuit of ecstasy there was no freedom to act/move differently, all things were interacting wholly, and automatically/instinctively, in accordance with their qualitative feelings. This situation changes when rationality/awareness is brought forth into being. Nevertheless, Solar-Systic unfoldment since the bringing forth of rationality/awareness could barely have been any different (we will soon explore exactly why this is so). This means that it was never going to be the case that an event in the past occurred which would have prevented the bringing forth of ecstasy. So, if one believes that an asteroid strike massive enough to wipe out life on Earth could have occurred in the past, thereby preventing the bringing forth of ecstasy, then one is mistaken. Similarly, if one believes that the asteroid that is associated with the extinction of the dinosaurs might have missed the Earth, then one is mistaken.

It might seem like there are a plethora of possible paths to the bringing forth of ecstasy, due to events such as asteroid strikes modifying the evolutionary trajectories of planetary life as it strives for ecstasy. However, in reality, this isn't the case. There is only one Solar-Systic path to the bringing forth of ecstasy.

The relationship between ecstasy & individual instantiations of ecstasy

It is time for us to explore the nature of the relationship between ecstasy and individual instantiations of ecstasy. It goes without saying that every individual instantiation of ecstasy is a thing of immense importance; after all, ecstasy is the most precious and valuable state that the Universe can be in. Ecstatic activity exists at particular distinct locations, but ecstasy really needs to be seen as a term that refers to a collective set of activities. It is this collective set of activities which has cosmic significance. One can visualise ecstasy as a singular diasporic entity which has an immense plethora of tendrils at particular distinct locations; these ecstatic tendrils/activities collectively constitute a thing, an exceedingly distinct object within 'the hierarchy of spatio-temporal embeddedness'. We have labelled this distinct object: 'the thing that is the hallmark of ecstasy'.

Ecstasy is the act of bringing forth and purposefully interacting with technology; in other words, ecstasy is the activity of the master modifier. When technology has been brought forth into existence it doesn't stay the preserve of an individual instantiation of ecstasy! Rather, technology pervades far and wide and every time that technology gets utilised, being used for the purpose for which it was designed, then ecstasy exists. It is not in the nature of ecstasy to be isolated and static; ecstasy spreads out, transforms, connects, develops and multiplies. It is the summation of all of these interactions with technology across the planet, the aggregate of all this activity, which constitutes ecstasy. It is this aggregate activity which is of importance to life on Earth and which has cosmic significance.

We need to keep in mind that ecstasy has two elements; firstly, the activities that bring forth technological devices into existence; secondly, the purposeful interactions with these devices. In terms of the first element, the human species is the zenith of the evolutionary progression of life in our Solar System because it is the master modifier; it is that which brings forth technological devices. In terms of the second element, that which purposefully interacts with technological devices is, by definition, part of ecstasy, part of the human species.

When we look at the state of life on Earth in order to assess how healthy it is looking, what shape it is in, in terms of the likelihood of it continuing to survive and thrive into the distant future, then there are two questions which need to be asked. Firstly: Has ecstasy evolved? Secondly: How far has ecstasy travelled along the journey that is the universal pursuit of ecstasy? Once ecstasy has been brought forth it goes through several stages of unfoldment. Firstly, it slowly pervades. Secondly, it proliferates and explodes, as it creates the epoch of the 'technological explosion'. Thirdly, it extends its reach and significance through the sufficient deployment of technology, the attainment of which is the purpose of ecstasy. Fourthly, it is tamed in the era of the 'spiritual explosion' as the force to environmental destruction comes into a state of perfect balance with the force to environmental sustainability. This process of unfoldment clearly highlights the fact that it is ecstasy as a collective/whole which is of cosmic importance.

The process of unfoldment that we have just been considering raises another issue which needs to be explored. This issue is the concept and reality of freedom. It is clear that ecstasy 'as a whole' lacks freedom. Ecstasy 'as a whole' can only develop in one direction. Once the first piece of technology has been brought into existence it is inevitable that it will lead to a globalised technological society, and to the four stages of unfoldment just outlined.

What about individual humans? In contrast to the lack of freedom of ecstasy 'as a whole', individual humans have freedom. In other words, individual humans possess the ability to choose how to act. An individual human even has the ability to choose whether or not to become an instantiation of ecstasy! To put this more accurately, a 'non-human biological human' has the ability to choose whether or not to become a human. This ability to choose is at its most clearest in the case of suicide. Despite our Solar System providing an individual human with everything that it needs to positively thrive, everything that it needs to reach the highest possible states of ecstasy, states of immensely preferable excitation/exhilaration, some humans choose to kill themselves! How terrible!! Individual humans have the freedom either to act in accordance with their inner qualitative feelings/wisdom, which is something that will make them happy/fulfilled, or to not do this and be unhappy/unfulfilled/miserable. We have seen that ecstasy pays a heavy price for being able to experience the immensely pleasant; it leaves itself vulnerable to the possibility of experiencing immensely unpleasant states of extreme misery, pain and discomfort. It is when one reaches the depths of miserableness that one can easily tip over the edge into suicide. What all of this means is that despite the existence of freedom, that there really isn't very much freedom! One has the freedom to be miserable and terribly unhappy, and the freedom to end one's existence, but if one wants a reasonably happy and fulfilling life then one has to act in a particular way, the way that makes one happy and fulfilled. And, of course, one cannot decide which actions make one happy and fulfilled! If one had this power, then one would have real freedom, rather than being a cosmic puppet!

So, whilst it is true that individual instantiations of ecstasy have freedom, they actually have much less freedom than one might suppose. Individual humans can defy the universal pursuit of ecstasy either by killing themselves or by being miserable. However, the overwhelming majority of humans prefer to be alive rather than dead, and prefer to be happy rather than miserable. And, thankfully, the vast majority of humans have at least an inkling as to how they need to act, what they need to do, in order to be happy; they need to try and act in accordance with their inner qualitative feelings/wisdom. As the vast majority of humans want to live and want to be happy, this means that the vast majority of humans will act in accordance with the qualitative feelings within them that are the universal pursuit of ecstasy. In other words, despite the freedom of individual humans, the direction of human culture, the preordained unfolding of the journey that is the universal pursuit of ecstasy, is never in any doubt.

Two views of individual human purpose

That the *human species = ecstasy* has a cosmic purpose has been the central theme of our ongoing exploration into the nature of the Universe. We need here to address the question of whether or not individual humans, individual instantiations of ecstasy, also have a purpose. Does every human have a role to play in achieving ecstasy's purpose? Or, do some humans have such a role/purpose, whilst others do not? Alternatively, does it make little sense to talk of individual humans having a purpose?

Given that the purpose of ecstasy is to bring forth the sufficient deployment of technology, we can easily make sense of the idea that the humans that are involved with the development, deployment and maintenance of the sufficient deployment of technology have a purpose. These humans are, in a real practical sense, fulfilling ecstasy's purpose. One of the issues that we are interested in here is the question of whether or not it makes sense to say that such a practical fulfilment is their unique purpose. Was this practical fulfilment always the destiny of these particular individuals? The broader issue that we are interested in is the question of whether every human has a purposeful part to play in the fulfilment of ecstasy's purpose.

After all, if individual humans do not have a purpose, then one might reasonably wonder how on earth it is possible for ecstasy's purpose to be fulfilled! It will be fruitful for us to consider an analogy between ecstasy and a football team. The purpose of a football team is to be victorious over their opposition. One can easily comprehend how this purpose can be achieved because each of the players in the team has an individual purpose. The goalkeeper has the purpose of stopping the ball from going into their net, the defenders each have the purpose of giving protection to the goalkeeper in a particular area of the pitch, the purpose of the midfielders is to control the game, whilst the strikers have the purpose of scoring goals. Because of the existence of these individual purposes it is easy to grasp how the purpose of the team can be fulfilled. If all, or most, members of the team fulfil their individual purpose to a high level then there is a good chance that the team will be victorious. However, if one considers the possibility that the team has the purpose of being victorious, but that the members of the team have no individual purpose, then it is much harder to imagine how the team's purpose can be fulfilled. If the goalkeeper does not have the purpose of stopping the other team scoring, then they might stand still as the opposing striker kicks the ball towards their goal. If the strikers do not have the purpose of scoring goals, then they might simply decide to take a nap on the pitch! Of course, it is possible that a team that is devoid of individual purpose might be victorious; but if this happened, then one would be rather puzzled as to how such an extremely unlikely outcome occurred.

Our Solar System is a simple thing, not a puzzling thing, so we would expect ecstasy's purpose to be achieved through the process of individual humans having purposes which they fulfil. Our question is, therefore: How can we make sense of the idea that individual humans have a purpose and that the summation of these individual purposes, summation across space, and summation across time, leads to the fulfilment of ecstasy's purpose? We need to envision that in order for ecstasy's purpose to be fulfilled that there are an enormous number of 'mini purposes' which need to be fulfilled. For example, the microchip needed to be invented. The invention of the microchip is a 'mini purpose' which plays a part in fulfilling ecstasy's purpose. This 'mini purpose' existed before the coming into being of the individual who in actuality fulfilled it. There is nothing mysterious about these 'mini purposes'. From the first moment that life on Earth arose, all of the steps which needed to be taken in order for life on Earth to attain the sufficient deployment of technology were clear. These steps are the 'mini purposes' that need to be fulfilled if ecstasy is to be brought into being and if ecstasy is to fulfil its purpose.

Let us consider some of these 'mini purposes'. One is 'the invention of the wheel', another is 'the initiation of the Scientific Revolution', another is 'the bringing forth of ecstasy', another is 'the invention of the telescope', another is 'the bringing forth of the steam engine', and another is 'the recognition of human-induced global warming'. These required steps, these 'mini purposes', need to be fulfilled if ecstasy's purpose is to be fulfilled. The process of fulfilling 'mini purposes' was occurring long before ecstasy came into existence, and the very bringing forth of the human species itself represents the fulfilling of a 'mini purpose'. Our main concern here is what happens after this bringing forth. In other words, we are concerned with the 'mini purposes' that need to be fulfilled in a particular part of the journey that is the universal pursuit of ecstasy, the part of the journey that is the realm of human cultural evolution. Let us consider two possibilities concerning how these 'mini purposes' come to be fulfilled.

Firstly, one could think of 'mini purposes' as 'empty vessels' which are waiting to be filled. It is because the steps which needed to be taken in order for life on Earth to attain the sufficient deployment of technology were clear at the moment that life arose, that it could be fruitful to think of these steps as 'empty vessels' which are waiting to be filled. From this perspective, the activities of ecstasy at any particular moment in time will be filling 'empty vessels' and thereby taking life on Earth further along its journey towards the sufficient deployment of technology. When all of the 'empty vessels' are filled the human species will have fulfilled its purpose. One important thing to appreciate is that on this 'empty vessel' view it matters not a jot which individual human fills a particular empty vessel. As long as some human, any human, fills an empty vessel, ecstasy will then be another step closer to the fulfilment of its purpose. What is important is the state of progression of ecstasy 'as a whole'; individual instantiations of ecstasy are dispensable cogs in the ecstatic machine. According to this way of thinking, there isn't one particular individual human whose purpose it was to bring forth the microchip. There is no individual destiny in this sense. It is also important to realise that it is possible for lots of humans to simultaneously fulfil the same 'mini purpose'; one example is: 'use an unsustainable amount of resources'.

The second possibility, the alternative to the 'empty vessel' view, is the 'unique destiny' view. This is the view that particular individuals have a unique purpose/destiny, a purpose/destiny which only they can fulfil. These two views are polar opposites – the 'empty vessel' view is focused on a 'mini purpose' which can be filled by any human, whilst the 'unique destiny' view is focused on a particular human and what their specific purpose is. According to the 'unique destiny' way of thinking, it was the unique purpose/destiny of the actual inventor of the microchip to be the inventor of the microchip.

The 'empty vessel' view is appealing because one might find it to be highly implausible that only one individual human has the specific purpose of inventing a particular thing, such as the microchip. What would have happened if this was so and the particular individual in question was the unfortunate subject of a terrible accident early in their life which resulted in their death? Would the microchip never have been invented? One surely wants to accept that the lives of humans who lived in the past could have turned out differently; one wants to accept that the actual human who invented the microchip could possibly have died at a young age and that if this happened that someone else would then have invented the microchip. If one believes this, then it obviously cannot be the case that it was the unique purpose, the destiny, of the actual human who invented the microchip to be the inventor of the microchip. In short, the 'unique destiny' view is intellectually unattractive because if the particular human in question for some reason did not fulfil their purpose – perhaps they had an untimely death, perhaps they were very unfortunate and succumbed to a debilitating disease, or perhaps they just didn't want to fulfil their potential having lapsed into a state of misery, then one surely still wants to think that the microchip would have been invented.

The first possibility that we have considered, the 'empty vessel' view, is 'the way that things are'. This means that particular individual humans are not essential to the fulfilment of ecstasy's purpose. If the particular human who actually fulfilled the 'mini purpose' of bringing forth the microchip had died before inventing the microchip, then another part of ecstasy would have fulfilled this 'mini purpose' because it is an 'empty vessel' that is waiting to be filled. We can now explore in much more detail, in the context of the 'empty vessel' view, what exactly it means to talk of individual human purpose.

Human potential & the filling of 'empty vessels'

It is time for us to explore how the vast majority of humans play a part in bringing about the fulfilment of ecstasy's purpose through the filling of 'empty vessels'. We have used 'inventing the microchip' as one example of a 'mini purpose'. There are obviously an immense number of 'mini purposes' that need to be fulfilled if ecstasy's purpose is to be fulfilled. And, of course, many of these 'mini purposes' have no possibility of being fulfilled until other 'mini purposes' have been fulfilled. One cannot fulfil the 'mini purpose' of 'inventing the mobile phone' until the 'mini purpose' of 'inventing the microchip' has been fulfilled. Sometimes the fulfilment of 'mini purposes' is a comparatively slow process, whilst at other times a flurry of 'mini purposes' are fulfilled in a very short period of time. This differential rate of 'mini purpose' fulfilment is, of course, closely related to the two forces in human cultural unfoldment that we explored earlier. The early twenty-first century is a period in which a plethora of 'mini purposes' are getting fulfilled at great speed. This isn't surprising given that we are living in the midst of the 'technological explosion'!

The questions which we need to address are: How do 'empty vessels' get to be filled? And, what is the mechanism through which humans come to fill 'empty vessels'? Without such a mechanism it is possible that there could be a complete mismatch between the 'empty vessels' that need to be filled and the actions and activities of the human species. We have actually already partially explored the mechanism through which human actions match and fill 'empty vessels'! We explored how individual humans have the freedom either to act in accordance with their inner qualitative feelings/wisdom, which will make them happy/fulfilled, or to not do this and be unhappy/unfulfilled/miserable. The mechanism of 'empty vessel' filling is simply that the vast majority of the human species acts in a way that makes itself happy, rather than in a way that makes itself miserable. Another way of putting this is to say that the mechanism involves humans fulfilling their potential to varying degrees of fullness; for, individual human potential is intimately connected to inner qualitative feelings/wisdom. If a particular human is utilising their abilities and talents, that is to say, if they are fulfilling their potential, then this will result in qualitative feelings that the human finds to be pleasant. If a human does the opposite, if they do not utilise their abilities/talents, if they do not fulfil their potential, then the result is unpleasant qualitative feelings; if cultivated, these feelings can lead to extreme misery and even suicide. So, in practice, acting in accordance with one's inner qualitative feelings/wisdom, fulfilling one's potential, means being aware of what one's abilities/talents are and utilising them as fully as one can.

Most of the 'empty vessels' that constitute the path of human cultural unfoldment represent different types of technological bringing forth and utilisation; however, there are many other 'empty vessels' in this path, such as those that represent the mental appreciation of particular facts. As the *human species = ecstasy* is the master modifier it is not surprising that humans, in the main, fulfil their potential by modifying things to varying degrees and in different ways. Such modification activities can involve bringing forth new technologies into existence or purposefully interacting with existing technologies. If the journey that is the universal pursuit of ecstasy was a ladder, 'the day of the realisation of cosmic purpose' would be at the top of the ladder, the unequally spaced rungs of the ladder would be 'empty vessels', and each time a human filled an 'empty vessel' this would equate to stepping up a rung of the ladder. This ladder would have an immense number of rungs, and whilst some human acts, such as the bringing forth of the microchip, would result in a large step up the ladder, other human acts, such as purposefully interacting with an existing technology, would result in a very small step up the ladder.

Let us ask ourselves these questions: Why does the process of particular humans fulfilling their potential result in the filling of 'empty vessels'? Why is there a close correspondence between the 'empty vessels' that need to be filled if ecstasy is to fulfil its purpose, and the activities that particular humans will naturally engage in because these activities are activities that involve exercising their talents and abilities in order to fulfil their potential and be happy? Such a close correspondence is not a coincidence! What a miracle it would be if it were a coincidence! Our Solar System has brought forth ecstasy into existence because it needed a particular set of abilities/talents if it was going to be able to sustain life. In other words, the abilities/talents that ecstasy has are designed for the job. The abilities/talents that the human species has are the abilities/talents that life on Earth needs in order to bring forth the sufficient deployment of technology. There is nothing surprising here! There are no miracles! Humans fulfilling their potential will automatically, and inevitably, result in the filling of 'empty vessels'.

Whilst the purpose of ecstasy is to bring forth the sufficient deployment of technology, not all 'empty vessels' within the realm of human culture are directly related to the utilisation and complexification of technology. Some 'empty vessels' that need to be filled are acts of mental appreciation that result from that aspect of the master modifier that is the drive to probe and understand through scientific and non-scientific enquiry. So, for example, ecstasy needs to come to appreciate the existence of the 'greenhouse effect', and it also needs to appreciate the existence of the phenomenon of human-induced global warming. Other 'empty vessels' that need to be filled in the realm of human culture have the purpose of keeping humans sane so that they are able to keep on working, keep on bringing forth new technological devices, keep on consuming and living, and keep on breeding. So, the progression of human culture needs 'empty vessels' such as 'comedy', 'art', 'music', 'religion', 'sport' and 'entertainment' to be filled. All of the abilities and talents that have been gifted to humans are valuable. Even the comedian is part of the journey that is the universal pursuit of ecstasy!

Fulfilling one's potential

How can we tell whether or not a particular human is fulfilling their potential? The first thing to say is that human potential varies greatly. In order to fulfil their potential some humans will need to make major breakthroughs which spread throughout human culture, whilst other humans will fulfil their potential through mundane day-to-day activities such as purposefully interacting with various technologies that have been brought forth by others. When one considers whether or not a particular human is fulfilling their potential one needs to consider factors such as the following: Is this human utilising their talents and abilities in the domain of which they are passionate? Has this human embraced life and sought to do that which makes them satisfied/happy? Does this human find their occupation and/or their life to be exhilarating? Is this human being all that they can be? Does this human glow and tingle with qualitative fulfilment when they are engaged in their work/passions/activities? Does this human have a 'sparkle' in their eye? If at least a couple of these questions have affirmative answers, then it is probably fair to say that this human is fulfilling their potential.

Fulfilling one's potential requires knowing oneself; one needs to know what one's talents and abilities are if one is to utilise them, if one is to become all that one can be. Furthermore, one needs to deploy these talents/abilities in accordance with one's passions/values. The human species is still learning how to do this effectively. Most humans could be fulfilling much more of their potential than they currently are, and this is because we are currently in the stage of our unfolding Solar System that is the age of the 'technological explosion'. The attainment of high levels of fulfilment of individual human potential will only increase dramatically throughout the human species when we enter the age of the 'spiritual explosion'.

It is easy to see how the development of human culture encourages individual humans to fulfil their potential, through the appeal of things such as recognition, promotion, financial reward, peer pressure, sports cars, amazing houses, perks, champagne, luxurious holidays, wonderful possessions, the attention of a desirable partner, and the opportunities that are presented for fulfilling one's potential. Those humans who do not fulfil their potential are more likely to be unsatisfied, miserable, lonely and financially poor. Having said this, it is, of course, possible to both fulfil one's potential and be financially poor. However, the fulfilment of potential naturally attracts financial reward, and the prospect of such a reward clearly helps to motivate a lot of humans to fulfil their potential. Our Solar System provides as many motivating factors as possible, because some humans are motivated by particular things, whilst other humans are motivated by very different things. Despite the immense variety of motivating factors, a very small minority of humans only fulfil a miniscule amount of their potential. How sad this is! This is very sad indeed, but our interest is ultimately not in individual humans, we are interested in the overall aggregate actions of ecstasy. And the vast majority of humans do fulfil their potential to varying degrees of fullness. How wonderful this is! The result of this wonderful fulfilment is the onward marching of life on Earth towards the sufficient deployment of technology.

As we have already explored, the 'mini purposes' that are available to be fulfilled will vary through time (so, one cannot invent the mobile phone if the microchip has not yet been invented). Particular humans are born at a particular stage of cultural unfoldment and this stage provides a certain range of possibilities to enable them to fulfil their potential. These possibilities are, of course, the 'empty vessels' that are available to be filled at this time. If a human fulfils their potential to a reasonable degree then they will automatically be filling some of these 'empty vessels'. So, all humans have a purpose; all humans have a role to play in the fulfilment of ecstasy's purpose. The purpose of individual humans is to fulfil their potential, to utilise their abilities and talents, to be happy and satisfied, and through so doing to fill 'empty vessels'. As the vast majority of individual humans fulfil their potential to varying degrees of fullness, they each play a part in fulfilling ecstasy's purpose.

There is one remaining issue which we should address. This is the question of why it is that different humans have different sets of skills/talents/abilities. There is no doubt that humans are often very different; some humans have a particular set of skills/talents/abilities, whilst other humans have very different sets of skills/talents/abilities. These immense differences are of great value, of crucial importance, because they ensure that the human species is provided with the diverse range of skills/talents/abilities that life on Earth needs in order to attain the sufficient deployment of technology. The question before us is how it comes to be that particular humans come to have particular sets of skills/talents/abilities. The answer to this question is to be found in the swirling patterns of unfoldment of our Solar System. For, the individual is forged by the state of the whole. If one is being formed and brought forth into existence at a certain stage of Solar-Systic unfoldment, then one will have a different set of skills/talents/abilities to those that one would have had if one was brought into existence at another stage of this swirling unfoldment. This is all that we need to appreciate. If you are keen to nail down the specifics of this process, and gain a proper appreciation of why you have the specific skills/talents/abilities that you do, then you would need to become an expert in astrology. However, unless attaining such an understanding is your passion, you don't need to do this. You just need to appreciate that your unique skills/talents/abilities are a gift from our Solar System, a gift that you should seek to utilise to the best of your ability.

In short, the universal pursuit of ecstasy is operating within us, flowing through us, through the movement to 'increasingly intense pleasantness' within us that arises when we utilise our skills/talents/abilities in the domain of that of which we are passionate. As instantiations of rationality/awareness we have the possibility of either aiding the universal pursuit of ecstasy (fulfilling our potential) or resisting it (barely fulfilling any of our potential, and maybe even completing suicide). So, the issue of the purpose of one's life can clearly be viewed from two perspectives, that of the individual and that of the whole of which the individual is a part.

Environmentalism

In the context of humans fulfilling their potential through filling 'empty vessels' it will be fruitful for us to explore the subject of environmentalism. Recall that there are two forces running through human culture – the force to environmental destruction and the force to environmental sustainability. These forces have been created through the past and current filling of particular 'empty vessels'. The future existence of these two forces requires the future filling of 'empty vessels'. Environmentalism is obviously part of the force to environmental sustainability. In other words, the actions undertaken by humans that fall under the domain of environmentalism are essential actions; these actions fill 'empty vessels' that need to be filled if life on Earth is to continue its ecstatic journey to the sufficient deployment of technology.

In order for the sufficient deployment of technology to be brought forth, the human belief that such a deployment is required needs to exist. This belief is partially generated by an increasing realisation of the enormous extent to which ecstasy has perturbed, and continues to perturb, the biogeochemical cycles of the Earth. This increasing realisation inevitably leads to increasingly environmentally-friendly actions, as ecstasy, in its guilt-tinged state, comes to terms with what it has done; these actions are themselves 'empty vessels' which need to be filled. The increasing realisation of the extent to which it has perturbed the biogeochemical cycles of the Earth culminates in the realisation that ecstasy needs to deploy technology to actively control the temperature of the Earth's atmosphere. Once this belief has taken root, then it is full steam ahead to the sufficient deployment of technology. The primary non-human-caused need for the sufficient deployment of technology – the impending collapse of life on Earth's first line of defence – only becomes widely appreciated after the coming into existence of the sufficient deployment of technology, when our Solar System has unfolded to the point that is 'the day of the realisation of cosmic purpose'.

There is perhaps something slightly paradoxical about all of this, for, most humans who are currently filling the 'empty vessels' of environmentalism would, no doubt, be horrified about the prospect of the sufficient deployment of technology! This is because there is still a strong belief system / story underpinning environmentalism which entails that the ideal situation is for humans to 'leave things alone as much as possible' and to 'stop meddling in nature'. This story is bought into by those who have been infected by the trap of the flawed extrapolation, which we explored earlier. Such a story gets its most extreme elucidation in the paradigm known as 'deep ecology'. This paradigm has its uses, but it arises out of a fundamentally flawed view of the relationship between ecstasy and the unfolding Solar System of which it is an immensely precious part.

So, environmentalism is an exceptionally good thing. It is fantastic that humans are utilising their skills/talents/abilities in the realm of environmentalism, through recycling, renewable energy, campaigning, writing and many other activities. Such activities fulfil 'mini purposes' and help to propel life on Earth forward towards both the sufficient deployment of technology and the age of the 'spiritual explosion'. In this epoch the force to environmental sustainability will become so strong that the human species will eventually live in perfect harmony with its surroundings and it will have immense respect for all life-forms. In other words, things will have come to a head and the two forces – the force to environmental sustainability and the force to environmental destruction – will have come into a state of perfect balance.

Embracing extreme uncertainty

The universal pursuit of ecstasy arises from the desire of things to attain 'increasingly intense pleasantness' through the ability to control, to modify, the other. The ability to control and modify the other reaches its zenith with the coming into existence of ecstasy. Ecstasy is the zenith of modification ability because it is able to utilise its penetrating ability to probe its surroundings through scientific investigation, and then use this in-depth knowledge to modify and control its surroundings in the most mind-bogglingly extravagant ways. The knowledge of its surroundings gained by ecstasy, and the ensuing ability to control those surroundings, is central to its very being; it is effectively its essence as ecstasy. This means that uncertainty is the antithesis of ecstasy. Ecstasy cannot stand uncertainty! In its ideal world ecstasy would understand everything, it would know everything, and it would be able to perfectly predict every movement that ever happens in its surroundings. The master modifier cannot bear the possibility that the Universe might be fundamentally unknowable and unpredictable, so it deludes itself into believing that it understands the Universe. This is where the phenomenon of 'human prejudice' comes into play; ecstasy convinces itself that the vast majority of the Universe is very different to itself, just a dead unfeeling mechanism which is governed by 'laws' and which is wholly explicable through science. Of course, as by now you are well aware, widespread human understanding of the actual nature of the Universe only comes into existence when the journey that is the universal pursuit of ecstasy has unfolded to the point that is 'the day of the realisation of cosmic purpose'.

In certain realms of human culture uncertainty is recognised and embraced; of course it is. For example, in the realm of environmentalism, the effort to embrace uncertainty in a sensible way has led to the formation of the 'precautionary principle', which entails that we should assume that the worst-case scenario will prevail and act accordingly. The problem is that due to its inevitably deluded belief that it understands more than it actually does, ecstasy consistently underestimates the level of uncertainty that actually exists. Sometimes the level of this underestimation is ginormous! In the realm of environmentalism, this results in a range of forecasts which ecstasy believes to be very diverse and broad in scope, but which are all actually in a very narrow and conservative range. For example, the IPCC has produced a range of forecasts concerning the possible level of the Earth's atmospheric temperature by the year 2100. This range of forecasts is formulated within a belief system which necessitates that ecstasy believes that it has more knowledge concerning the nature of the Universe than it actually has. What this means is that even the most pessimistic forecasts could be overshot, and not just by a small margin, but by an enormous margin!

If we could fully comprehend the extent of the uncertainty that exists then this would be a good thing. Embracing extreme uncertainty leads to the realisation that there is an extreme need for a vastly increased ability to control and modify. Such an increased ability to control and modify provides the means through which a high level of future uncertainty can be adequately dealt with by ecstasy for the benefit of life on Earth. In other words, embracing extreme uncertainty paves the way for the speedy bringing forth of various technological protectors of life on Earth, chiefly the sufficient deployment of technology.

To what extent can we change the future?

The existence of the universal pursuit of ecstasy means that large swathes of the unfolding of our Solar System are predetermined. The overwhelming majority of the Universe is devoid of rationality/awareness, which means that the way that these parts of the Universe interact with each other is wholly determined by the nature of the qualitative feeling states involved. The way that these qualitative feeling states interact with each other is a wholly automatic process – states that involve 'increasingly intense pleasantness' are 'held onto' whilst states that are unpleasant are 'pushed away'. This automatic process leads to the unfolding of the Universe; this is a process we can look back on and call the 'evolution' of the Universe. These automatic interactions might even be interpreted from without as being in obeyance of 'laws' due to their regularity.

What this means is that before the bringing forth of rationality/awareness everything that happened in the Universe, every movement, couldn't have been any different. The coming into existence of rationality/awareness in the Universe was an event of momentous significance; this is because rationality/awareness has the ability to choose. Some of the instantiations of rationality/awareness in the Universe are humans. We have already explored the ability to choose alternative courses of action that humans have. Individual humans can choose to take heed of their qualitative feelings/wisdom, and the more that they take heed, the more that they will fulfil their potential. Alternatively, individual humans can choose to ignore their qualitative feelings/wisdom, with the result that they will become miserable to varying degrees, with suicide being the end-point of extreme misery.

Humans do have freedom, but they are also 'cosmic puppets'. Individual humans have the ability to not act in accordance with their inner qualitative feelings/wisdom. However, if one chooses to not act in accordance with the qualitative feelings/wisdom within one, if one rebukes the universal pursuit of ecstasy, if one doesn't utilise any of one's skills, abilities and talents, then one will be miserable. This is why humans are 'cosmic puppets'; the freedom that exists is rather limited in its nature. Let us imagine that a stranger comes up to you and says: "I will give you a choice. I can either put my hand in my right pocket, take out my gun and shoot you in the head; or, I can put my hand in my left pocket, take out my wallet and give you a thousand pounds. You choose." In this situation you obviously have the freedom to make a choice. But how much freedom is there here really!! There will undoubtedly be the occasional oddball who in this situation will think: *what the hell, I might as well die,* just like there are occasional humans who complete suicide. However, for the overwhelming majority of people there is no choice here, there is only one way in which they can possibly act. This is a rather stark example, but it nicely exemplifies the human situation as 'cosmic puppets'.

We have seen that a human has the freedom to choose whether or not to live a happy life that involves fulfilling their potential. But who would choose not to live a happy life?! Are there humans who really want to be miserable? Surely not! Of course not! The reason that humans become miserable is that it isn't a simple endeavour to truly know oneself, to know how to fulfil one's potential, to properly take heed of one's inner qualitative feelings/wisdom, and to thereby become happy. This is why there are battalions of life-coaches, counsellors, careers advisors, psychoanalysts, spiritual gurus and psychologists, all of whom seek to enable people to take control of their own lives and to fulfil their potential.

Human actions are overwhelmingly driven and determined by the universal pursuit of ecstasy, which exists in the human body as a multitude of qualitative feelings. It is inevitable that the vast majority of humans will take heed of these feelings, due to the potency of the natural desire for 'increasingly intense pleasantness'. This means that the vast majority of humans will fulfil their potential to varying degrees of fullness. In turn, this means that the direction of human culture, from the bringing forth of ecstasy, to the 'technological explosion', to the 'spiritual explosion', is predetermined. In other words, it is inevitable that ecstasy will bring forth the sufficient deployment of technology. The question that we are currently exploring is: To what extent, if any, can we change the future? It goes without saying, given what has already been said, that there is very little scope for us to change the future. Yet, we do have freedom! We have freedom as individuals, yet this freedom evaporates when we move from the scale of the individual to human culture as a whole.

It will be fruitful for us to ruminate upon the following question: Is it possible to intentionally change the future unfoldment of human culture? Human culture is currently in the midst of the age of the 'technological explosion'; as this epoch continues to unfold it will reach its victorious crescendo with the fulfilment of ecstasy's purpose: the bringing into being of the sufficient deployment of technology. This glorious event will be followed by the age of the 'spiritual explosion' which culminates in 'the day of the realisation of cosmic purpose'. When we reach 'the day of the realisation of cosmic purpose' we will be truly free; 'cosmic puppets' no more! However, given the widespread understanding of human cosmic purpose that exists following this realisation, the expression of true freedom will not be something that leads to anything radical, such as the destroying of the wonderful technological fruits that ecstasy has been working so tirelessly to bring into being! We don't have any control over any of this; we cannot stop it happening; we are the puppets, not the puppet master! Is there any possibility for us to change the future in a meaningful way? Whilst we cannot change the direction in which human culture is heading, an issue that presents itself is whether there is any scope for us to speed up the rate of its unfoldment.

Even if this were possible, why would we want to speed up the rate at which human culture unfolds? The age of the 'technological explosion' is the age in which suffering in life on Earth – the suffering of ecstasy and the suffering of non-ecstatic life-forms – is maximised. If we could, ever so slightly, speed up our movement out of the age of the 'technological explosion', then we would be able to reduce the amount of suffering experienced by life on Earth. How could we attempt to do this? We could simply fulfil our cosmic purpose as soon as possible. It is the bringing forth of the sufficient deployment of technology that heralds the end of the era of the 'technological explosion'. You are well aware that the widespread realisation that the purpose of the human species was to bring forth the sufficient deployment of technology is an event which follows this deployment; it occurs at the zenith of the age of the 'spiritual explosion' on 'the day of the realisation of cosmic purpose'.

We are currently, in the year 2021, gradually drifting in ignorance towards the sufficient deployment of technology. However, if a small group of humans became acutely aware that the sufficient deployment of technology is an inevitability, that it is a wonderful thing for life in our Solar System, and that it is urgently needed, then it is possible to imagine that this group might be able to lead a conscious effort to push full steam ahead towards the sufficient deployment of technology. In this way, human cultural unfolding could seemingly be speeded up a little. However, in reality, such an outcome is extremely hard to conceive because there would be fierce resistance to such a move from the vast majority of humans, who inevitably, given their present delusion, cannot see the need for the sufficient deployment of technology. The inevitable existence of this fierce resistance implies that we lack the ability to speed up the future unfoldment of human culture. Despite having rationality/awareness, the human species seems to be powerless to speed up the unfolding journey that is the universal pursuit of ecstasy.

Nevertheless, it will be fruitful for us to further ponder the positive outcomes that would arise if we were able to speed up the future unfoldment of human culture. We are currently wasting an enormous amount of resources due to the widespread erroneous belief that reducing fossil fuel emissions is a solution to the global warming situation that we face. If there was an increasing realisation that the effort to reduce emissions in this way is a 'red herring', then these resources could be immediately reallocated to the technologies that are deployed and sufficiently deployed. However, it is certainly possible that we don't have any ability to change the future unfoldment of human culture in any meaningful way! Puppets. Puppets. Puppets. *Things will unfold at the pace that they are meant to unfold at.* It seems likely that we will continue to waste resources on the emissions reduction 'red herring' until we are forced to reluctantly turn to the technologies that are deployed and sufficiently deployed. This delay, this continued focus on reducing fossil fuel emissions, will cause immense avoidable death and suffering. But, unfortunately, it seems as if such death and suffering are needed in order to get us off our backsides, to propel us out of our state of deluded lethargy and into action, to power us forward towards the technologies that are deployed and sufficiently deployed.

However, there is perhaps another possible future; a more desirable future; to explore this possible future you can read the section: *The "Mysterious Prince".* You never know, perhaps this possible future, this more desirable future, will come into being. Perhaps this possible future is the pace that things, the pace that human culture, was always meant to unfold at! How wonderful this would be, if it were to come to pass.

If one has had a glimpse into the wonderful future that awaits life on Earth then it is tempting to believe, to hope, that this future can be speedily ushered in both so that one can experience it, and so that the suffering of the life-forms of the Earth can be minimised. But this is just wishful thinking. One has to accept that one was born at a particular stage of our unfolding Solar System and engage positively with the reality that this entails. The wonderful future is for those who are yet to be born!

Whilst it is extremely doubtful that the rate of the future unfoldment of human culture can be speeded up, individual humans can definitely change their future. Individual humans can increasingly get to know themselves, and in so doing they can increasingly fulfil their potential and enhance their level of happiness/fulfilment. In tandem with this, individual humans can also come to realise, and accept, that they are a cosmic puppet; and, through such acceptance mental suffering and anguish can be extinguished. So, the future can be changed. If a puppet comes face-to-face with its master, then its future can be one of increasing happiness and peace.

And, of course, if we attempted to speed up the rate of human cultural unfoldment, we would still be left facing these questions: Did we actually, through our conscious intent, speed up the rate at which human culture unfolded? Or, was this always the pace at which human culture was meant to unfold at?

The feminisation of our Solar System

There is one aspect of the unfoldment of human culture that is worthy of elucidation. You will be well aware that the dominant driver of the evolution of human culture, from its beginnings to the current day, is the force that we have called the force to environmental destruction. This force is the unfettered desire to modify which dominates the vast majority of the journey that is the universal pursuit of ecstasy. As human culture unfolds, a second force – the force to environmental sustainability – grows in strength as ecstasy rationalises that the force is of immense positive value. This force can be thought of as a necessary constraint on the force to environmental destruction. These two forces, acting in tandem, lead to the fulfilment of ecstasy's purpose: the bringing into being of the sufficient deployment of technology.

You might recall that we discussed how these two forces could quite aptly be referred to by various labels. Another way of referring to these forces, as they manifest themselves in human culture, would be to talk of the 'masculine' force (the force to environmental destruction) and the 'feminine' force (the force to environmental sustainability). To talk of two forces running through human culture which are 'masculine' and 'feminine' is simply to recognise that the dominant driver of the evolution of human culture, from its beginnings to the current day, is a force which has attributes that are aptly referred to by the human concept of the 'masculine'. Whilst, as human culture progresses, as the Solar-Systic journey that is the universal pursuit of ecstasy reaches a certain stage in its unfoldment, then an opposing force, a force which has attributes that can aptly be referred to by the human concept of the 'feminine', gradually becomes much more powerful and significant. The 'masculine' force dominates the unfoldment of human culture throughout the age of the 'technological explosion', but as this age progresses the 'feminine' force gradually increases in strength. When our Solar System unfolds to the age of the 'spiritual explosion', it will then become progressively feminised. At the zenith of the age of the 'spiritual explosion', on 'the day of the realisation of cosmic purpose', human culture will be sufficiently feminised that the two forces will come into a state of perfect balance. More than this, at this stage of the unfoldment of human culture, the two forces will become one.

Cherishing the living

Our Solar System has brought forth life. It is alive! It is buzzing with life! It has brought into being a plethora of wonderful life-forms; that is to say, it has brought forth a plethora of amazingly fantastic states of heightened excitement/exhilaration! It has even brought forth ecstasy!! How wonderful all of this is! Life is rare. Life is very rare. Most of the Universe is barren. Most of the Universe is depressed and lifeless. The universal pursuit of ecstasy is everywhere – the entire Universe is perennially striving to bring forth and maintain life; however, the vast majority of the time its efforts are thwarted, the conditions are just not right in most of the Universe for life to be brought forth into existence.

The Earth is the womb of life in our Solar System. Earthly conditions have enabled life to evolve and to positively thrive. The universal pursuit of ecstasy has been successful; it has flourished. Ecstasy exists! We exist!! Let us fully appreciate the rareness of life. Let us cherish all of the life-forms that currently exist on the Earth; every one of these life-forms is a minor miracle, a rarity, an extremely precious part of our Solar System. We are all brothers/sisters in striving, we are all striving to survive through striving to modify to the best of our ability. When an individual life-form dies, this is a cause for sadness; when it survives, this is a cause for celebration. We might be the master modifier, the zenith of the evolutionary progression of life on Earth, the saviour of life on Earth, the most precious life-form in our Solar System and in any solar system, but we don't have to see our fellow planetary life-forms as inferior, as resources to be used and eaten. We needed to see our fellow planetary life-forms this way in the past, we needed to use and abuse our fellow planetary life-forms in the past, but there is certainly no need to do so now. Let us cherish each and every one of these immensely precious things.

Paradise on Earth

We are currently living through a period of immense turmoil within life on Earth; turmoil for ecstasy itself and turmoil for the non-ecstatic part of planetary life. However, the Earth has a wonderful future. 'The day of the realisation of cosmic purpose' will occur at the zenith of the age of the 'spiritual explosion', and at this time it will be sublimely wonderful to be a life-form on the Earth. At this point in the unfoldment of our Solar System paradise will come into existence on the Earth.

At this future time ecstasy will be spiritually enlightened and it will not act in an exploitative way towards any life-form. The period of turmoil which enabled the bringing forth of the technological fruits which are, in this paradise, enabling life on Earth to survive and thrive, is now a distant memory. This paradise is not a return to the past; it is a harmonious synthesis of super-advanced technology and enlightened ecstatic being. The super-advanced technology is dominated by the sufficient deployment of technology; however, many other life-protecting technological fruits also exist. Enlightened ecstatic being involves ecstasy living sustainably, respectfully and peacefully. There is a simple, sustainable, respectful, harmonious lifestyle embedded within the super-advanced technology that enables the thriving of life on Earth.

There will be a time in the extremely distant future when the Earth becomes uninhabitable. Our focus throughout our exploration of the universal pursuit of ecstasy has been on the immediate and short-term future of our planet and Solar System. In the very near future, the sufficient deployment of technology will be brought forth and it will enable the coming into being of paradise on Earth for an immensely long stretch of time. However, in the very distant future, when the Earth becomes uninhabitable due to the near-expiration of the Sun, we can be certain that the life that has arisen in the womb of our Solar System will have escaped to another Solar System to continue its glorious existence!

Celebration

Wouldn't it be nice if we could celebrate our existence? We are ecstasy! We are the saviours of life on Earth! The survival of our non-human brothers/sisters is our responsibility; we need to take care of them through bringing forth the sufficient deployment of technology. Oh, if we could see all this, we could then celebrate our existence! Oh, if we could realise how important and special we are! Oh, if we could see the cosmic responsibility that rests on our shoulders! Could we possibly bear to come face-to-face with this immense responsibility? Our

Solar System is celebrating us; it is joyous at our arrival. Yet, we are miserable! We are in a state of despair! We inevitably tell ourselves that our species is selfish and greedy and that our presence on the Earth is destructive and harmful. We even tell ourselves that we are the enemy of our non-human brothers/sisters. How deceived we are. How wonderfully deceived!

One day humans will be celebrating. One day will be 'the day of the realisation of cosmic purpose'. On this day, every human will appreciate the value of all humans, all cultures, all perspectives and all life-forms. On this day, every human will see the nature of human cosmic purpose; they will see the Solar-Systic journey that is the universal pursuit of ecstasy laid out before them in its entirety. The wonderful universal pursuit of ecstasy!

Our starting point

"What is man? – so I might begin; how does it happen that the world contains such a thing, which ferments like a chaos or moulders like a rotten tree, and never grows to ripeness? How can Nature tolerate this sour grape among her sweet clusters?"

Our destination

And so it ends – the human species is ecstasy; the world contains such a chaotic and mouldering thing because only such a thing can save life on Earth from obliteration. The sweet clusters of life would turn to dust if it were not for this sour grape! The human species does grow to ripeness, but it can only do so when it has fulfilled its cosmic purpose. The joyous arrival of the technological fruit that is the sufficient deployment of technology enables the transformation of this sour grape into the epitome of sweetness.

Gaian Wisdom, Astrology

&

Scientific Knowledge

Gaian Wisdom

James Lovelock:

"We may at first think that there is nothing particularly odd about this picture of a stable climate over the past three and a half eons [3,500 million years]... Yet it is odd, and for this reason: our sun, being a typical star, has evolved according to a standard and well established pattern. A consequence of this is that during the three and a half aeons of life's existence on the Earth, the sun's output of energy will have increased by twenty-five per cent."

(Gaia: A New Look at Life on Earth, OUP, 2000, p. 18)

"right from the beginning of life, around three and a half aeons ago, the Earth's mean surface temperature has never varied by more than a few degrees from its current levels. It has never been too hot or cold for life to survive on our planet, in spite of drastic changes in the composition of the early atmosphere and variations in the sun's output of energy."

(Gaia: A New Look at Life on Earth, OUP, 2000, p. 48)

"If our planetary temperature depended only on the abiological constraints set by the sun's output and the heat balance of the Earth's atmosphere and surface, then... all life would have been eliminated."

(Gaia: A New Look at Life on Earth, OUP, 2000, p. 20)

"only for a brief period in the Earth's history was the sun's warmth ideal for life, and that was about two billion years ago. Before this it was too cold for comfort and afterwards it has progressively grown too hot... The sun is already too hot for comfort."

(*The Revenge of Gaia*, Penguin Books, 2006, p. 44-5)

"[Gaia] is old and has not very long to live. As the sun grows ever hotter it will, in Gaia's terms, soon become too hot for animals and plants and many of the microbial forms of life."

(*The Revenge of Gaia*, Penguin Books, 2006, p. 46)

"The brief interglacials, like now, are, I think, examples of temporary failures of ice-age regulation."

(*The Revenge of Gaia*, Penguin Books, 2006, p. 45)

Stephan Harding:

"There are tipping points beyond which climate can suddenly transmute from benign to deadly, and there is no good reason for us to bask in the complacent idea that our emissions of greenhouse gases will warm the planet gradually – that we will have time to adapt. It is far more likely that we will trigger abrupt, catastrophic climate changes that will push Gaia into a new hot state unsuitable for many of her life forms, including ourselves."

(*Animate Earth: Science, Intuition and Gaia,* Green Books, 2006, p. 188)

Astrology

Richard Tarnas:

"The current body of accumulated data makes it difficult to sustain the modern assumption that the universe as a whole is best understood as a blind, mechanistic phenomenon of ultimately random processes with which human consciousness is fundamentally incoherent, and in which the Earth and human beings are ultimately peripheral and insignificant. The evidence suggests rather that the cosmos is intrinsically meaningful to and coherent with human consciousness; that the Earth is a significant focal point of this meaning, a moving center of cosmic meaning in an evolving universe, as is each individual human being; that time is not only quantitative but qualitative in character, and that different periods of time are informed by tangibly different archetypal dynamics; and, finally, that the cosmos as a living whole appears to be informed by some kind of pervasive creative intelligence – an intelligence, judging by the data, of scarcely conceivable power, complexity, and aesthetic subtlety, yet one with which human intelligence is intimately connected, and in which it can consciously participate. I believe that a widespread understanding of the potent but usually unconscious archetypal dynamics that coincide with planetary cycles and alignments, both in individual lives and in the historical process, can play a crucial role in the positive unfolding of our collective future."

(*Cosmos and Psyche,* Plume, 2007, p. 489)

Scientific Knowledge

IPCC:

"Each of the last four decades has been successively warmer than any decade that preceded it since 1850. Global surface temperature in the first two decades of the 21st century (2001-2020) was 0.99 [0.84-1.10] °C higher than 1850-1900. Global surface temperature was 1.09 [0.95 to 1.20] °C higher in 2011-2020 than 1850-1900."

(IPCC, 2021: Summary for Policymakers. In: *Climate Change 2021: The Physical Science Basis. Contribution of Working Group I to the Sixth Assessment of the Intergovernmental Panel on Climate Change. Cambridge University Press:* A.1.2)

Stephen Peake & Joe Smith:

"Even if we engage in further actions to reduce GHG emissions, the models tell us that anthropogenic climate change will continue for centuries."

('A citizen's guide to climate science'. In Peake, S., and Smith, J. eds. 2003. *Climate Change: From Science to Sustainability.* Milton Keynes: The Open University. p. 78)

Robert H. Stewart:

"New CO_2 is released into the atmosphere when fossil fuels and trees are burned. Roughly half of the CO_2 released into the atmosphere quickly dissolves in the cold waters of the ocean which carry it into the abyss...therefore temporarily reducing atmospheric CO_2. Eventually, however, most of the CO_2 must be released back."

(*Introduction to Physical Oceanography.* Florida: Orange Grove Texts Plus. 2009. p. 214, p. 233)

IPCC:

> "Human influence is *very likely* the main driver of the global retreat of glaciers since the 1990s and the decrease in Arctic sea ice area between 1979-1988 and 2010-2019 (about 40% in September and about 10% in March)...It is *very likely* that human influence has contributed to the observed surface melting of the Greenland Ice Sheet over the past two decades."

> (IPCC, 2021: Summary for Policymakers. In: *Climate Change 2021: The Physical Science Basis. Contribution of Working Group I to the Sixth Assessment of the Intergovernmental Panel on Climate Change. Cambridge University Press:* A.1.5)

L. Caesar, *et al:*

> "Here, we compare a variety of published proxy records to reconstruct the evolution of the AMOC[3] since about AD 400. A fairly consistent picture of the AMOC emerges: after a long and relatively stable period, there was an initial weakening starting in the nineteenth century, followed by a second, more rapid, decline in the mid-twentieth century, leading to the weakest state of the AMOC occurring in recent decades."

> ('Current Atlantic Meridional Overturning Circulation weakest in last millennium', *Nature Geoscience,* Vol. 14, March 2021, pp. 118-120)

---

[3] AMOC stands for 'Atlantic Meridional Overturning Circulation'. The AMOC is part of the thermohaline circulation. From our perspective, the perspective of the universal pursuit of ecstasy, we can think of the term AMOC as referring to the thermohaline circulation.

Thomas L. Delworth, *et al:*

"It is very likely that the strength of the AMOC will decrease over the course of the 21st century in response to increasing greenhouse gases, with a best estimate decrease of 25-30%."

('The Potential for Abrupt Change in the Atlantic Meridional Overturning Circulation', In Delworth, T. L. (2008), *Abrupt Climate Change.* Reston: U.S. Geological Survey. p. 259)

Jorge Sarmiento & Corinne Le Quere:

"the oceanic uptake of atmospheric carbon dioxide was substantially reduced in scenarios involving global warming relative to control scenarios. The primary reason for the reduced uptake was the weakening or collapse of the ocean thermohaline circulation."

('Oceanic Carbon Dioxide Uptake in a Model of Century-Scale Global Warming', *Science,* Vol. 274, No. 5291, Nov. 22, 1996, pp. 1346-1350)

Lijing Cheng, *et al:*

"The most recent data indicate that the ocean heat content (OHC) in the upper 2000 m layer of the world's oceans has increased with a mean rate of 5.7+/- 1.0 Zetta Joules yr^{-1} for the 1958-2020 period. There is a more rapid increase in OHC that began ~1980s and has continued unabated since then."

('Upper Ocean Temperatures Hit Record High in 2020', *Advances in Atmospheric Sciences,* Vol. 38, April 2021, pp. 523-530)

NASA:

"Is the ocean continuing to warm? Yes, the ocean is continuing to warm. Notably, all ocean basins have been experiencing significant warming since 1998, with more heat being transferred deeper into the ocean since 1990. To date, the ocean contains 90 percent of the heat from human-induced global warming, and the year 2020 was the warmest ever measured for the global ocean."

(NASA, 2021, Available at: https://climate.nasa.gov/faq/53/is-the-ocean-continuing-to-warm [Accessed: 28 August 2021])

Harold F. Upton & Peter Folger:

"All gases, such as CO_2, are less soluble in water as temperature increases."

('Ocean Acidification', *Congressional Research Service,* July 30, 2013, p. 2)

Friedrich Hölderlin

& the

Environmental Crisis

The German Romantic Friedrich Hölderlin lived from 1770 to 1843. He developed a unique perspective on the relationship between humankind and the rest of nature. He believed that humanity has a positive role to play in cosmic evolution, and that modernity is the crucial stage in fulfilling this role. In this section I will be arguing for a reinterpretation of his ideas regarding the position of humankind in cosmic evolution, and for an application of these ideas to the 'environmental crisis' of modernity. This reinterpretation is of interest because it entails an inversion of the conventional notion of causality in the 'environmental crisis'; instead of humans 'harming' nature, in the reinterpretation it is nature that causes human suffering.

Hölderlin's ideas are of particular interest because he yearned for an end to human suffering, but was also firmly convinced that humankind was inevitably destined to be separated from nature, and thereby destined to endure suffering. Hölderlin's conception of the human-nature relationship as part of an unfolding process of cosmological change seems to be of great relevance today, an age that is characterized by belief in the meaninglessness of human existence, and by concern about the way that we have altered the pre-human conditions of the Earth. Hölderlin's views provide a unique perspective on modernity that is worthy of serious consideration.

I start by outlining Hölderlin's views on the role of humankind in universal evolution. I then review the secondary literature on Hölderlin that relates to these ideas. I proceed to argue that Hölderlin's philosophy is applicable to, and gives a unique perspective on, the 'environmental crisis' of modernity. I argue that the existing secondary literature on Hölderlin has not recognized this, and that a reinterpretation of the role of humanity in Hölderlin's philosophy of cosmic evolution is therefore required. My central claim is that for Hölderlin, modernity and the related notion of the contemporary 'environmental crisis' is a necessary stage of cosmic evolution, and thus that it is far from a 'crisis'. Rather, it is a necessary stage of disharmony that will inevitably be followed by a re-conquered harmony. I will argue that this disharmony is characterized by the environmental changes that are resultant from the development of technology.

1. Hölderlin's philosophy of human nature, cosmic evolution and modernity

The starting point of Hölderlin's philosophy is that there must be a basic unknowable reality which precedes self-consciousness wherein subjects and objects are not in existence but are both part of a 'blessed unity of being'. He describes this unity as, "Where subject and object simply are, and not just partially, united...only there and nowhere else can there be talk of being."[4] He argues that the 'blessed unity of being' (which he also refers to as 'nature') is responsible for the coming into existence of humanity through using its power to initiate a division of itself into subjects and objects. This division of being causes the emergence of judgement. Hölderlin states that, "'I am I' is the most fitting example of this concept of judgement...[as] it sets itself in opposition to the *not-I,* not in opposition to *itself.*"[5]

The division means that human beings are not capable of actions that are independent of nature; Hölderlin states that, "all the streams of human activity have their source in nature."[6] It is revealing to compare this claim with the words of Hölderlin's character Hyperion, "What is man? – so I might begin; how does it happen that the world contains such a thing, which ferments like a chaos or moulders like a rotten tree, and never grows to ripeness? How can Nature tolerate this sour grape among her sweet clusters?"[7] For Hölderlin, man is the 'violent' being, whose coming into existence in opposition to the rest of nature was *initiated* by nature.

Hölderlin sees this opposition between man and the rest of nature as culminating in modernity – an era that he claims is characterised by the absence of the gods. In *Brot und Wein* Hölderlin writes, "Though the gods are living, Over our heads they live, up in a different world...Little they seem to care whether we live or do not."[8] A key question for Hölderlin is how we deal with this separation. He envisions two possibilities – the 'Greek' response which is to dissolve the self and die, and the 'Hesperian' response of a living death.

Hölderlin came to view the 'Greek' response as hubristic, it being based on an anthropocentric (human-centered) desire to oppose the division initiated by nature. He thus sees the 'Hesperian' response of living and carrying out actions that are dependent on nature for their origination as the appropriate non-hubristic response to our separation. Hölderlin's position is that as nature created the separation, *only* nature can bring the separation to an end. He sees this process of separation and reconnection as part of a broader cosmic picture wherein nature is an unfolding organism rather than a huge mechanism. This organismic view enables him to envision teleological processes in nature which give rise to his claim that there will be, "eternal progress of nature towards perfection."[9]

---

[4]Friedrich Hölderlin, 'Being Judgement Possibility', in J. M. Bernstein (ed.), *Classic and Romantic German Aesthetics,* Cambridge, Cambridge University Press, 2003, p. 191.
[5]Ibid., p. 192.
[6]Alison Stone, 'Irigaray and Hölderlin on the Relation Between Nature and Culture', in *Continental Philosophy Review,* vol. 36, no. 4, 2003, p. 423.
[7]Friedrich Hölderlin,'Hyperion', in Eric L. Santner (ed.), *Hyperion and Selected Poems,* New York, Continuum, 1990, p. 35.
[8]Ibid., p. 185.
[9]Ronald Peacock, *Hölderlin,* London, Methuen & Co. Ltd, 1938, p. 36.

2. Interpretations of Hölderlin and his concept of fate

In this section I set out my view of Hölderlin's conception of fate – that all human actions are part of the evolution of nature towards perfection. I do this by reviewing the existing scholarly literature on Hölderlin and showing that whilst these interpretations all recognise parts of Hölderlin's conception of fate that they do not capture the whole of it. I start with interpretations of human nature, move on to cosmic processes, and finally consider the role of modernity within these processes.

At the level of the human there is a general consensus in the literature that Hölderlin's position is that humans are endowed by nature with qualities that shape human nature, and that this inevitably shapes human interactions with the rest of nature. There are various names in the literature for the qualities which are endowed to humans. Dennis J. Schmidt refers to the qualities present in humans as their 'formative drive.' He claims that, "Hölderlin suggests that human nature and practices are to be understood by reference to a formative drive which expresses itself as a constant need for 'art'."[10] In a similar vein, Thomas Pfau argues for an 'intellectual intuition.' He states that, "Hölderlin recasts the convergence of "freedom and necessity" as the most primordial synthesis of intellect and intuition itself, a synthesis which takes place within the subject itself. He thus approaches what Kant had repeatedly ruled out as an "intellectual intuition"."[11]

In agreement with Schmidt and Pfau, Franz Gabriel Nauen argues that for Hölderlin, "all men do in fact have the same basic character...all human activity can be derived from the same *elemental drive* in human nature."[12] The 'formative drive' / 'intellectual intuition' / 'elemental drive' identified in the literature reveals why man can be seen as the 'violent' being. Human nature is to engage in 'art', to utilize the resources of nature so that culture can be generated and sustained. This generation of human culture actually benefits nature as a whole, but it requires large-scale modification of parts of non-human nature. The destiny of man is thus a disruptive one. It is clear that it is also an undesirable one. Nauen states that for Hölderlin, "Even war and economic enterprise serve to fulfil the destiny of man, which is to "multiply, propel, distinguish and mix together the life of Nature"."[13]

So Hölderlin sees human nature, economic production and even war as parts of a broader cosmic evolutionary process; the Universe *as a whole* is seen as evolving to perfection. There will inevitably be aspects of this evolution that from a narrow perspective could be viewed as 'less than perfect'. These negative aspects of the evolutionary process – from war, to the presence of evil in its entirety – have to be seen as inescapable parts of the whole process.

The key point is that for Hölderlin the cosmic evolutionary process *ends* in perfection. Thus, Ronald Peacock argues that, "the division produced by conflict is followed by a re-conquered

---

[10]Dennis J. Schmidt, *On Germans and Other Greeks,* Indiana University Press, 2001, p. 139.

[11]Thomas Pfau, *Friedrich Hölderlin: Essays and Letters on Theory,* New York, SUNY Press, 1988, p. 15.

[12]Franz Gabriel Nauen, *Revolution, Idealism and Human Freedom: Schelling, Hölderlin and Hegel and the Crisis of Early German Idealism,* Indiana University Press, 2001, p. 139.
[13]Ibid.

harmony."[14] Similarly, Anselm Haverkamp argues that an interpretation of the poems *Andenken* and *Mnemosyne* is the expression, 'where danger threatens, salvation also grows.'[15] Whilst, Martin Heidegger translates the opening lines of *Patmos* as, "But where danger is, grows the saving power also."[16] Hölderlin's view is clearly that from a narrow and short-term perspective danger and conflict are often the norm, but that these things actually play a part in bringing about a greater harmony in the future. In the long-term they are all part of the evolution of the whole Universe to perfection.

Cosmic evolution is thus one long process of disharmonies and inevitably following harmonies. Peacock argues that Hölderlin's vision is of a, "harmonised process of life which comprises within itself the rhythmic movement from chaos to form and back again, and an emotional experience of this which in the sphere of nature knows only the one rapture, but in the human sphere suffering and joy."[17] It is revealing that this interpretation sees 'violent' humans as suffering, whilst nature is purely rapturous. This clearly sheds light on the question posed by Hölderlin's character Hyperion: "How can Nature tolerate this sour grape among her sweet clusters?"[18] The answer seems to be that human 'violence' *enables* nature to be rapturous. And, as part of this rapture, humans experience suffering.

Why should suffering be a uniquely human experience? To explain this Peacock cites part of a letter from Hölderlin to his brother, "Why can they [humans] not live contented like the beasts of the field? he asks: and replies that this would be as unnatural in man, as in animals the tricks, or arts, man trains them to perform. Thus he establishes that the arts of man are natural to man. Culture, then, derives from nature; and the impulse to it is the characteristic which distinguishes man from the rest of creation."[19]

The human impulse to culture has culminated in the era of modernity. Hölderlin sees this period as one of great significance as he sees it as a historical epoch that is characterised by the *absence of the gods*. To be consistent with his views on harmonised evolution to perfection there must be a reason for this absence. Indeed, Peacock argues that Hölderlin thinks that, "a godless age is part of a divine mystery, it is as necessary as day, ordained by a higher power."[20] Furthermore, Heidegger claims that the gods are still present, despite their absence: "man who, even with his most exulted thought could hardly penetrate to their Being, even though, with the same grandeur as at all time, they were somehow there."[21]

The absence of the gods in modernity is deeply related to the contemporary danger that exists in modernity. It should be remembered that this danger cannot be a cause for concern for

[14]Peacock, *Hölderlin*, p. 22.

[15]Anselm Haverkamp, *Leaves of Mourning: Hölderlin's Late Work,* New York, SUNY Press, 1996, p. 48.

[16]Martin Heidegger, 'The Question Concerning Technology', in R.C. Scharff and V. Dusek (eds.), *Philosophy of Technology: The Technological Condition – An Anthology,* Oxford, Blackwell Publishing, 2003, p. 261.

[17]Peacock, *Hölderlin*, p. 22.

[18]Hölderlin, 'Hyperion', p. 35.

[19]Peacock, *Hölderlin*, p. 36.

[20]Ibid., p. 92.

[21]Martin Heidegger, *Existence and Being,* London, Vision Press Ltd., 1956, p. 190.

Hölderlin – as all dangers are inevitably followed by regained harmonies. Nevertheless, Heidegger attempts to identify the exact danger that Hölderlin believed is present in modernity. Heidegger claims that, "the essence of technology, enframing, is the extreme danger."[22] It must follow that for Heidegger, "precisely the essence of technology must harbor in itself the growth of the saving power."[23] He sees this as occurring when the essential unfolding of technology gives rise to the possibility of opening up a "free relation" with technology which is inclusive of non-instrumental possibilities.[24]

In an interpretation of the 1802 hymn *Friedensfeier*, Richard Unger draws out Hölderlin's views on the absence of the gods in modernity.[25] In *Friedensfeier* the entire span of Western civilization is characterised as a thunderstorm which is ruled by a "law of destiny" which ensures that a certain amount of "work" is achieved. Unger argues that it is clear that this "work", "is the product of the storm itself and that it designates the harmonious totality of earthly existence during the coming era."[26] The end of the "storm" of modernity enables the arrival of a mysterious "prince" who makes it possible that, "men can now for the first time hear the "work" that has been long in preparation "from morning until evening"."[27]

Following the inevitable successful accomplishment of the "work" of Western civilization, the great Spirit will disclose a Time-Image which will, "be a comprehensive depiction of the historical process and its triumphant result."[28] Unger argues that, "the Image shows that there is an alliance between the Spirit of history and the elemental divine presences of nature – for the natural elements with which man has always worked have played integral and essential parts in man's history."[29] The triumphant result of the actions of humankind in modernity is clearly an example of a re-conquered harmony that follows division.

In Unger's interpretation of *Friedensfeier* we have a picture of modernity in which humans are carrying out "work" under a "law of destiny". The crucial factor is that humanity is ignorant that it is working under a "law of destiny" in modernity, until modernity has ended. It is then that through the Time-Image the great Spirit reveals the successful outcome of modernity, and the *nature and value* of the accomplished "work". This is a prime example of a short-term and narrow perspective entailing the perception of a lack of destiny and of needless suffering, whilst in the longer-term the same events are seen to be an inevitable part of a broader positive outcome – the evolution of the Universe to perfection.

This difference of perspectives can explain an apparent contradiction in the literature between Unger's interpretation of *Friedensfeier,* and Schmidt's analysis of Hölderlin's 1801 letter to

---

[22]Heidegger, 'The Question Concerning Technology', p. 261.

[23]Ibid.

[24]R.C. Scharff and V. Dusek, 'Introduction to Heidegger on Technology', in R.C. Scharff and V. Dusek (eds.), *Philosophy of Technology: The Technological Condition – An Anthology,* Oxford, Blackwell Publishing, 2003, p. 248.

[25]Richard Unger, *Friedrich Hölderlin,* Boston, Twayne Publishers, 1984, pp. 100-105.

[26]Ibid., p. 102.

[27]Ibid., p. 101.

[28]Ibid., p. 104.

[29]Ibid., p. 105.

Bohlendorff. This letter was written only one year before *Friedensfeier* and Schmidt claims that in it Hölderlin's position is, "that the peculiar flow of modernity is the lack of destiny."[30] The apparently contradictory views of Unger and Schmidt can be reconciled through recalling Peacock's interpretation that, "a godless age is part of a divine mystery, it is as necessary as day, ordained by a higher power,"[31] and comparing it to Unger's claim that men are blind to the point of the "work" that they have been carrying out until the "storm" of Western civilization has passed.

The comparison reveals that the "law of destiny" applies to the activities of *humanity as a collective* in Western history, activities that are ordained by a higher power for a specific purpose. In contrast, the "lack of destiny" applies to *individual human beings*. This difference arises because individual humans are unaware that their actions are part of an inevitably unfolding cosmic plan, it is only the fruition of the plan that enables realization. Instead, humans believe that they have free will and live in a meaningless age. Therefore, modernity can at one and the same time be characterized as both a period governed by a "law of destiny" and a period constituted by a "lack of destiny". The difference is purely one of perspective.

This conception of modernity as simultaneously being a period of a "lack of destiny" and a "law of destiny" raises the issue of human attitudes to nature. If human attitudes and actions towards nature are in the interests of nature, then it seems that there is no such thing as a truly 'human-centered' attitude. The appropriate attitude that humans should take to the objective side of nature, given Hölderlin's philosophy, has been addressed by Alison Stone. She argues that because, "according to Hölderlin's thinking, we have become separated from nature by *its* power alone, so it is not within *our* power to undo separation."[32] Therefore, "the appropriately modest response is to endure separation – to wait, patiently, until nature may change its mode of being."[33] This means that the appropriate human attitude entails, "the *acceptance* of disenchantment, of separation, of meaninglessness."[34]

This view is concordant with the "lack of destiny" perspective. However, when the "law of destiny" is taken into account, then the hidden meaning is revealed. Furthermore, the whole notion of the attitudes of individual humans then becomes irrelevant. It seems that there cannot be such a thing as a *truly* human-centered attitude, because all attitudes originate from nature, and they all lead to actions which fulfil the "law of destiny". It may seem that our attitudes to nature are of importance, but this is because we believe in a "lack of destiny", and are inevitably blind to the bigger picture of the "law of destiny". Whatever our attitudes as individuals, our relationship with the rest of nature as a collective would be 'for the best'.

---

[30]Schmidt, *On Germans and Other Greeks,* p. 137.

[31]Peacock, *Hölderlin,* p. 92.

[32]Stone, 'Irigaray and Hölderlin on the Relation Between Nature and Culture', p. 424.

[33]Ibid.

[34]Alison Stone, *Nature in Continental Philosophy – Week 4, Section V, Friedrich Hölderlin,* [online], http://www.lancaster.ac.uk/depts/philosophy/awaymave/408new/wk4.htm, [accessed 25 October 2005].

3. A reinterpretation of the human in cosmic evolution

The interpretations of Hölderlin that I have reviewed all give an accurate representation of Hölderlin's views. However, they are all partial views. They all miss the 'big picture' of what Hölderlin's views imply about what it means to be a human in the context of cosmic evolution, and the consequent implications for the perspective from which we should view modernity and the 'environmental crisis'. In an attempt to fully grasp these implications I am going to defend the thesis that: *Hölderlin's philosophy leads to the conclusion that the 'environmental crisis' is a necessary stage in the purposeful evolution of nature towards perfection*. This is an interesting thesis because, if accepted, it would supplant the conception of the meaninglessness of human existence with a conception of positive cosmic purpose.

The argument I will be making centers on three key aspects of Hölderlin's philosophy. Firstly, that he believes that nature is purposefully evolving towards perfection. Secondly, that he believes that the achievement of this perfection requires human actions. Thirdly, that he believes that human actions are determined by nature. Acceptance of these three claims leads to the conclusion that human actions are determined by nature as a necessary stage in the purposeful evolution of nature towards perfection. As the 'environmental crisis' of modernity is purely resultant from human actions, a second conclusion inevitably follows. This is that the 'environmental crisis' itself is determined by nature as a necessary stage in the purposeful evolution of nature towards perfection.

I will now present evidence to support the three key claims. The first claim is that Hölderlin's belief is that *nature is purposefully evolving towards perfection*. The Universe can either be viewed as a giant mechanism or as an unfolding organism; Hölderlin clearly held the latter view. This conception of the Universe explains his belief that nature unfolds in a way that serves its own purposes; that disharmonies are followed by regained harmonies. This is why Peacock claims that Hölderlin believed in, "the eternal progress of nature towards perfection,"[35] and, "the emergence of perfection in the course of natural development."[36]

This firm belief clashed with Hölderlin's personal yearning for immediate perfection in life. His immense desire to see a morally just world was completely at odds with his philosophical belief that the perfection he sought could only be achieved in the course of natural development. The movement to perfection envisioned by Hölderlin is thus a fatalistic one, an inevitable evolutionary progression towards perfection. Peacock captures this with his claim that for Hölderlin there is an, "acute sense of 'Fate', of inevitability, expressed again and again in his work. Fate is revealed in the process of history... it is inherent in the passage of form to chaos, and of disintegration to a new harmony."[37]

This first claim is the most straightforward of the three. The second claim is that *Hölderlin believes that the achievement of perfection requires human actions*. The starting point in

---

[35]Peacock, *Hölderlin,* p. 36.
[36]Ibid., p. 105.
[37]Ibid., p. 93.

defending this claim is Hölderlin's central belief that nature *used its power* to divide itself and thereby create humankind. This division means that the split was part of the evolutionary process rather than a random occurrence. We can ask ourselves why this split may have been a necessary occurrence. An initial answer seems to be Nauen's claim that, "Even war and economic enterprise serve to fulfil the destiny of man, which is to "multiply, propel, distinguish and mix together the life of Nature"."[38]

In *The Perspective from which we Have to look at Antiquity* Hölderlin asserts that, "antiquity appears altogether opposed to our primordeal drive which is bent on forming the unformed, to perfect the primordial-natural so that man, who is born for art, will naturally take to what is raw, uneducated, childlike rather than to a formed material where there has already been pre-formed [what] he wishes to form."[39] In a letter to his brother he also asserts that, "the impulse to art and culture…is really a service that men render nature."[40]

The source of Hölderlin's primordeal drive to art is nature, because it is nature that created us and endowed us with our capabilities. This is clear from Peacock's interpretation that, "Man cannot be master of nature; his arts, *necessary though they may be in the scheme of things,* cannot produce the substance which they mould and transform; they can only develop the creative force, which in itself is eternal and not their work."[41]

Hölderlin's primordeal drive to art in humans has inevitably led to the epoch of modernity. Human actions in this epoch appear to be central to the achievement of perfection. Hölderlin claims that modernity is an epoch that, "is as necessary as day, ordained by a higher power."[42] Furthermore, humans have been involved in "work" in modernity that is clearly constitutive of the importance of the epoch. This is clear from Unger's interpretation of *Friedensfeier* in which the "law of destiny" ensures that a certain amount of human "work" is done. The crucial factor is that humanity is ignorant that it is working under a "law of destiny" in modernity, until modernity has ended. It is then that through the Time-Image the great Spirit reveals the successful outcome of modernity, and the nature and value of the accomplished "work".

There is no doubt that in Hölderlin's view human actions and their resultant "work" in modernity are part of purposeful evolution to perfection. What is interesting is the exact nature of the "work". There is an obvious connection between the "work" of modernity (*Friedensfeier*) and the "danger" we face in modernity (*Patmos*). Heidegger's interpretation of *Patmos* that, "the essence of technology, enframing, is the extreme danger,"[43] makes it clear that the "work" of modernity is the development of technology. In fact, technological development in modernity seems to be the culmination of Hölderlin's primordeal drive to art.

---

[38]Nauen, *Revolution, Idealism and Human Freedom: Schelling, Hölderlin and Hegel and the Crisis of Early German Idealism,* p. 139.

[39]Friedrich Hölderlin, 'The Perspective from which We Have to Look at Antiquity', in Thomas Pfau (ed.), *Friedrich Hölderlin: Essays and Letters on Theory,* New York, SUNY Press, 1988, p. 39.

[40]Peacock, *Hölderlin,* p. 37.

[41]Ibid.

[42]Ibid., p. 92.

[43]Heidegger, 'The Question Concerning Technology', p. 261.

Furthermore, it is very hard to think of any other distinctive aspects of modernity that are resultant from human actions, present an extreme danger, and have cosmic significance. Therefore, for Hölderlin, the achievement of perfection requires the human development of technology.

It is interesting that Heidegger sees the danger we face from the "work" of modernity as the essence of technology rather than actual technology. Andrew Feenberg has criticised Heidegger for this abstract concentration on essences rather than the actual technology itself.[44] A 'Feenberg interpretation' of *Patmos* seems to me to be more in accordance with Hölderlin's views than the 'Heidegger interpretation', as Hölderlin's philosophy is grounded in actualities rather than essences. Hölderlin sees a positive role for actual technology in cosmic evolution; this means that *actual technology* has a cosmic purpose. Therefore, it seems that both the danger we face, and the saviour, must be the *actual* technology developed by human actions.

The importance of the human split from the rest of nature can also be seen in the words of Hölderlin's character *Hyperion*: "How should I escape from the union that binds all things together? We part only to be more intimately one, more divinely at peace with all, with each other. We die that we may live."[45] Human actions are thus depicted as a 'living death' that is necessary for the life (and continued movement to perfection) of nature as a whole. This explains Peacock's interpretation that, "the sphere of nature knows only the one rapture, but in the human sphere [there is] suffering and joy."[46]

The third claim is that *Hölderlin believes that human actions are determined by nature.* There are many passages in Hölderlin's novel *Hyperion* that attribute the responsibilities for human actions to a power or god: "There is a god in us who guides destiny as if it were a river of water, and all things are his element."[47]..... "oh forgive me, when I am compelled! I do not choose; I do not reflect. There is a power in me, and I know not if it is myself that drives me to this step."[48]..... "I once saw a child put out its hand to catch the moonlight; but the light went calmly on its way. So do we stand trying to hold back everchanging Fate. Oh, that it were possible but to watch it as peacefully and meditatively as we do the circling stars."[49]..... "Man can change nothing and the light of life comes and departs as it will."[50]..... "We speak of our hearts, of our plans, as if they were ours; yet there is a power outside of us that tosses us here and there as it pleases until it lays us in the grave, and of which we know not where it comes nor where it is bound."[51]

---

[44]Andrew Feenberg, 'Critical Evaluation of Heidegger and Borgmann', in R.C. Scharff and V. Dusek (eds.), *Philosophy of Technology: The Technological Condition – An Anthology,* Oxford, Blackwell Publishing, 2003, pp. 327-337.

[45]Hölderlin, 'Hyperion', p. 123.

[46]Peacock, *Hölderlin,* p. 22.

[47]Hölderlin, 'Hyperion', p. 11.

[48]Ibid., p. 79.

[49]Ibid., p. 22.

[50]Ibid., p. 127.

[51]Ibid., p. 29.

Hölderlin's belief in the lack of human free will is perhaps clearest in his claim in a letter to his mother regarding the views of Spinoza that, "one *must* arrive at his ideas if one wants to explain everything."[52] Spinoza's ideas can be summed up as, "Nature in all its aspects is governed by necessary laws, and human being no less than the rest of nature is determined in all its actions and passions, contrary to those who conceive of it as 'a dominion within a dominion'."[53]

In order to make abundantly clear Spinoza's – and thus Hölderlin's – views on a lack of human free will here are two quotes from Spinoza: "I say that thing is free which exists and acts solely from the necessity of its own nature...I do not place Freedom in free decision, but in free necessity."[54] And, "a stone receives from an external cause, which impels it, a certain quantity of motion, with which it will afterwards necessarily continue to move...Next, conceive, if you please, that the stone while it continues in motion thinks, and knows that it is striving as much as possible to continue in motion. Surely this stone, inasmuch as it is conscious only of its own effort, and is far from indifferent, will believe that it is completely free, and that it continues in motion for no other reason than because it wants to. And such is the human freedom which all men boast that they possess, and which consists solely in this, that men are conscious of their desire, and ignorant of the causes by which they are determined."[55]

Furthermore, in an interpretation of Hölderlin's *Stutgard,* Peacock argues that, "the laws of growth govern the culture as well as the lives of men...the one process comprehends all things and the one rhythm manifests itself again and again...in the progress of history; in the spiritual life of individuals."[56] In this vision not only human nature, but also the evolution of culture, is seen as an inevitable historical progression. Peacock's interpretation of Hölderlin is that, "man's spirit is but part of the One Spirit,"[57] which Hölderlin insists is involved in a "movement...through successive historical generations."[58] The spirit of man is thus governed by the larger Spirit of nature. This is the sense in which, "all the streams of human activity have their source in nature."[59]

The nature of the relationship between man's spirit and the Spirit of nature is made clear in the following quote from Hölderlin's character Diotima: "a *unique destiny* bore you away to solitude of spirit as waters are borne to mountain peaks."[60] This concept of individual humans having a unique destiny was the view of Johann Herder, who was one of Hölderlin's

---

[52]Friedrich Hölderlin, 'No.41: To his Mother', in Thomas Pfau (ed.), *Friedrich Hölderlin: Essays and Letters on Theory,* New York, SUNY Press, 1988, p. 120.

[53]Moira Gatens, *Imaginary Bodies: Ethics, Power and Corporeality,* London, Routledge, 1996, p. 111.

[54]Benedict de Spinoza, 'LVIII: To Schuller', trans. A. Wolf (ed.), *The Correspondence of Spinoza,* 2nd ed., London, Frank Cass & Co. Ltd., 1966, pp. 294-5.

[55]Ibid., p. 295.

[56]Peacock, *Hölderlin,* p. 25.

[57]Ibid., p. 90.

[58]Ibid., p. 114.

[59]Stone, 'Irigaray and Hölderlin on the Relation Between Nature and Culture', p. 423.

[60]Hölderlin, 'Hyperion', p. 122.

inspirations. Herder saw nature as a great current of sympathy running through all things which manifested itself in unique inner impulses within different individuals. This means that every human has a unique calling – an original path which they ought to tread. As Herder states, "Each human being has his own measure, as it were an accord peculiar to him of all his feelings to each other."[61] Clearly, for both Herder and Hölderlin, human actions at any one time are determined in accordance with the movements of the One Spirit of nature.

I have presented evidence for the claims that for Hölderlin: *nature is purposefully evolving towards perfection, the achievement of this perfection requires human actions, and human actions are determined by nature.* Acceptance of these three claims leads to the conclusion that human actions are determined by nature as a necessary stage in the purposeful evolution of nature towards perfection. I now briefly argue that the 'environmental crisis' of modernity is purely resultant from human actions.

The definition of an environmental problem is: "any change of state in the physical environment which is brought about by human interference with the physical environment, and has effects which society deems unacceptable in the light of its shared norms."[62] This definition encapsulates a sliding scale of environmental problems from those that are local and temporary, to those that are global and long-lasting. The 'environmental crisis' as a concept has arisen because of the emergence in the last 100 years of an increasing number of environmental problems that are towards the global and long-lasting end of the scale. The 'environmental crisis' is thus purely resultant from the *human actions* which have created environmental problems that are characterised by their global reach and long-lasting nature.

This means that the above conclusion, that human actions are determined by nature as a necessary stage in the purposeful evolution of nature towards perfection, needs amending. As the 'environmental crisis' is purely resultant from human actions, it too must be part of this purposeful evolution. Therefore, the new conclusion that inevitably follows is:

- *The 'environmental crisis' is determined by nature as a necessary stage in the purposeful evolution of nature towards perfection.*

4. Objections to the reinterpretation

It could be objected that there are many references to human freedom in Hölderlin's work that would seem to cast doubt on the third claim. This is particularly noticeable in his novel Hyperion. For example, Hyperion states that, "without freedom all is dead."[63] However, this

[61]Charles Taylor, *Sources of the Self: The Making of the Modern Identity,* Massachusetts, Harvard University Press, 1994, p. 375.
[62]Peter B. Sloep and Maris C.E. van Dam-Mieras, 'Science on Environmental Problems', in P. Glasbergen and A. Blowers (eds.) *Environmental Policy in an International Context: Perspectives,* Oxford, Butterworth-Heinmann, 2003, p. 42.
[63]Hölderlin, 'Hyperion', p. 117.

objection is easily answered because these references all appear in Hölderlin's early work, and even then they are more than counterbalanced by the opposing fatalistic views that I have outlined. In his early period Hölderlin was struggling to come to terms with the conflict between his keen moral aspirations for social change on the one hand, and his belief in perfection only arising through natural development on the other. In his later work, as is clear in his endorsement of the 'Hesperian' response to our condition, he firmly accepts the powers of natural development and the determination of human actions by nature. He realizes the futility of pursuing his idealistic moral aspirations because he accepts the illusory nature of human free will.

A further objection could be made that this reinterpretation is pointless because Darwin's theory of evolution, which emerged shortly after Hölderlin's time, gives a view of evolutionary processes that is incompatible with Hölderlin's view that there was a 'blessed unity of being' prior to the arrival of humans. We now know that the emergence of the human species – and its primordeal drive to art – was preceded by approximately four billion years of evolution of life on Earth. It can thus be argued that there was not a 'blessed unity of being' prior to the evolution of humankind.

This is exemplified by the claim of Hans Jonas that the subject-object divide opened up around four billion years ago, when, "living substance, by some original act of segregation, has taken itself out of the general integration of things in the physical context, set itself over against the world, and introduced the tension of "to be or not to be" into the neutral assuredness of existence."[64] This certainly does not appear to be a pre-human 'blessed unity of being'. However, it is interesting that Jonas also sees humans as, "a 'coming to itself' of original substance."[65]

It is clear that this Darwinian based objection does not invalidate Hölderlin's views, or the reinterpretation of them presented here. In fact, not only does evolutionary theory perfectly complement Hölderlin's philosophy, his philosophy *needs* it. The idea that nature could use its power to instantaneously create a being as complex as a human out of the 'blessed unity of being' is hardly defensible. In the light of our knowledge today we can simply reinterpret Hölderlin as claiming that nature used its power approximately four billion years ago to divide the 'blessed unity of being' and create a subject/object divide. As he sees nature as an unfolding and evolving organism, the divide would give rise to human subjects after a sufficient period of time. This, " 'coming to itself' of original substance", as Jonas describes it, has in actuality taken approximately four billion years.

---

[64]Hans Jonas, *The Phenomenon of Life: Toward a Philosophical Biology,* Illinois, Northwestern University Press, 2001, p. 4.
[65]Ibid., p. xv.

5. *Conclusion*

I have argued that the existing secondary literature has not grasped the full implications of Hölderlin's thought for what it means to be a human in modernity. By drawing together Hölderlin's ideas I have sought to understand his notion of the purpose of human actions, and what this purpose means for the 'environmental crisis'.

Hölderlin's conception of nature is an organism unfolding to perfection. I have argued that he sees modernity as an important stage of this unfolding, which is characterized by the development of technology through human actions. I have further argued that this means that the 'environmental crisis' of modernity – a side-effect of the development of technology – is also an inevitable stage of this unfolding; it is in the interests of nature. As nature continues to unfold, the disharmony of modernity will be succeeded by a re-conquered harmony. I have argued that Hölderlin's 'saving power' is actual technology, as this seems most consistent with his thought. Heidegger's view, that the 'saving power' is the essencing of technology, seems inconsistent with the positive role of technology in cosmic evolution that is envisioned by Hölderlin.

The reinterpretation I have outlined clearly entails an inversion of the conventional notion of causality in the 'environmental crisis' of modernity. Humanity is conventionally pictured as harming nature. My thesis has shown that for Hölderlin it is nature that is 'harming' humanity. We have been cast aside out of the rapture of nature into a realm of suffering and self-consciousness, with the purpose of developing technology to serve the purposes of the unfolding nature of which we are a part.

Finally, we are left with the question of what our attitudes to nature should be, given this reinterpretation of what it means to be a human in cosmic evolution. The answer is simple. As nature is the source of our individual attitudes, our attitudes to nature must be in the interests of nature. Our attitudes, whether they are techno-centric, environmentalist, quietist, or nature-exploitative are all correct for us as individuals, because in the aggregate they fulfil the purpose of nature as a whole.

Bibliography

Feenberg, Andrew, 'Critical Evaluation of Heidegger and Borgmann', in R.C. Scharff and V. Dusek (eds.), *Philosophy of Technology: The Technological Condition – An Anthology*, Oxford, Blackwell Publishing, 2003.

Gatens, Moira, *Imaginary Bodies: Ethics, Power and Corporeality*, London, Routledge, 1996. Haverkamp, Anselm, *Leaves of Mourning: Hölderlin's Late Work*, New York, SUNY Press, 1996.

Heidegger, Martin, *Existence and Being*, London, Vision Press Ltd., 1956.

Heidegger, Martin, 'The Question Concerning Technology', in R.C. Scharff and V. Dusek (eds.), *Philosophy of Technology: The Technological Condition – An Anthology*, Oxford, Blackwell Publishing, 2003.

Hölderlin, Friedrich, 'Being Judgement Possibility', in J. M. Bernstein (ed.), *Classic and Romantic German Aesthetics*, Cambridge, Cambridge University Press, 2003.

Hölderlin, Friedrich, 'Hyperion', in Eric L. Santner (ed.), *Hyperion and Selected Poems*, New York, Continuum, 1990.

Hölderlin, Friedrich, 'No.41: To his Mother', in Thomas Pfau (ed.), *Friedrich Hölderlin: Essays and Letters on Theory*, New York, SUNY Press, 1988.

Hölderlin, Friedrich, 'The Perspective from which We Have to Look at Antiquity', in Thomas Pfau (ed.), *Friedrich Hölderlin: Essays and Letters on Theory*, New York, SUNY Press, 1988.

Jonas, Hans, *The Phenomenon of Life: Toward a Philosophical Biology*, Illinois, Northwestern University Press, 2001.

Nauen, Franz Gabriel, *Revolution, Idealism and Human Freedom: Schelling, Hölderlin and Hegel and the Crisis of Early German Idealism*, Indiana University Press, 2001.

Peacock, Ronald, *Hölderlin*, London, Methuen & Co. Ltd, 1938.

Pfau, Thomas, *Friedrich Hölderlin: Essays and Letters on Theory*, New York, SUNY Press, 1988.

Scharff, R. C., and Dusek, V., 'Introduction to Heidegger on Technology', in R.C. Scharff and V. Dusek (eds.), *Philosophy of Technology: The Technological Condition – An Anthology*, Oxford, Blackwell Publishing, 2003.

Schmidt, Dennis J., *On Germans and Other Greeks,* Indiana University Press, 2001.

Sloep, Peter B., and Dam-Mieras, Maris C.E. van, 'Science on Environmental Problems', in P. Glasbergen and A. Blowers (eds.) *Environmental Policy in an International Context: Perspectives,* Oxford, Butterworth-Heinmann, 2003.

Spinoza, Benedict de, 'LVIII: To Schuller', trans. A. Wolf (ed.), *The Correspondence of Spinoza,* 2nd ed., London, Frank Cass & Co. Ltd., 1966.

Stone, Alison, 'Irigaray and Hölderlin on the Relation Between Nature and Culture', in *Continental Philosophy Review,* vol. 36, no. 4, 2003.

Stone, Alison, *Nature in Continental Philosophy – Week 4, Section V, Friedrich Hölderlin,* [online], http://www.lancaster.ac.uk/depts/philosophy/awaymave/408new/wk4.htm [accessed 25 October 2005].

Taylor, Charles, *Sources of the Self: The Making of the Modern Identity,* Massachusetts, Harvard University Press, 1994.

Unger, Richard, *Friedrich Hölderlin,* Boston, Twayne Publishers, 1984.

The "Mysterious Prince"

Hölderlin's "Mysterious Prince"

I have claimed that Hölderlin believed that technology has a positive role to play in cosmic evolution, and that whilst it presents a 'danger', it is fundamentally a 'saving power'. Given that Hölderlin lived before the human species formulated the concept of the 'environmental crisis', before there was an awareness of non-human-induced global warming, before there was an awareness of human-induced global warming, one cannot reasonably expect him to have known why technology is simultaneously a 'danger' and a 'saving power'.

Technology is a 'danger' because its unleashing results in immense planetary transformation and disruption – rainforests are replaced by cities, concrete, mining, roads and industrial agriculture, at the same time as tremendous amounts of fossil fuels are released from safe underground storage thereby creating an immense force for human-induced global warming. There are two elements to this immensely dangerous force. Firstly, this force for human-induced global warming leads to immediate actual global warming, which contributes to the ongoing planetary transformation through causing the warming of the oceans, the melting of icebergs and glaciers, and an increasing frequency of extreme weather events, such as floods, fires, storms and droughts. Secondly, this force has a pent-up element due to a ginormous amount of carbon dioxide getting sucked into the depths of the Earth's oceans where it goes on a journey of several centuries. Whilst it is on this journey, the world above goes through an immense transformation. This transformation is a process of weakening planetary robustness in the realm of the ability of life on Earth to keep its atmospheric temperature down – both the land surfaces and the oceans become less favourable homes for carbon dioxide. When the pent-up force comes to fruition, with the long-travelling oceanic carbon dioxide making its dramatic return into a vastly transformed world, a potent force for immediate global warming will come into being as carbon dioxide concentrations in the atmosphere spiral upwards.

The technologically-transformed planet is ill-equipped to deal with the immense force for human-induced global warming, because it was in an exceptionally weak state *before* it was weakened further by its technological transformation. It is the overwhelming of life on Earth's force for global cooling by the force for non-human-induced global warming which is the bringer of doom; whilst, the technological transformation of the planet is the penultimate nail in the coffin for life on Earth. This means that the birthing of technology brings with it a perfect storm of danger – a danger to the human species, and a danger to life on Earth.

Technology is fundamentally a 'saving power' because it is the 'armour' which life on Earth needs in order to survive and thrive into the distant future. The central component of this 'technological armour' is the technology which will be deployed to regulate the temperature of the Earth's atmosphere, through creating a force for global cooling. This deployment, this technological force for global cooling, is required because our Solar System is an unfolding whole which heats up as it ages. This one-way force for non-human-induced global warming causes the inevitable weakening, and expiration, of the non-technological force for global cooling that life on Earth deploys in the pre-technological era. Technology is life on Earth's 'saving power'. In the absence of technology life on Earth would be doomed, and would be hurtling towards obliteration.

In short, the process of bringing forth life on Earth's 'saving power', involves the creation of a period of immense disruption which entails the creation of an imminent 'danger' to the continued flourishing, and the continued survival, of life on Earth. The destiny, the cosmic purpose, of the human species is to bring into being life on Earth's technological 'saving power'. In order to fulfil its destiny, the human species needs to come face-to-face with the imminent 'danger' that is generated in the process of bringing forth life on Earth's 'saving power'. The human species can then save life on Earth through bringing forth a technological force for global cooling.

As a final reflection, it is worth recalling Hölderlin's belief that:

The end of the "storm" of modernity enables the arrival of a "mysterious prince" who makes it possible that men can now for the first time hear the "work" that has been long in preparation "from morning until evening".

Could the "storm" of modernity have ended?

Could this book enable humans to "hear" for the first time the purpose of their work?

Could I be Hölderlin's "mysterious prince"?!

Technology, Environmentalism

& the

Nature of the Universe

Human Dominion & the Nature of the Universe

There are two broad ways in which one can view the Universe. It is not an easy task to convey these ways adequately using words, but let us try. Firstly, one can view the Universe as a clockwork mechanism, which is devoid of intrinsic value, feeling and purpose. Secondly, one can view the Universe as an organism that is pervaded with intrinsic value – a feeling, striving and purposeful whole. There is little doubt that if one goes far enough back in time, back into human history, that the latter view, the view that the Universe is a feeling/striving/purposeful whole, was widespread and dominant. Whereas, the former view, the view that the Universe is a clockwork mechanism, has become increasingly widespread and dominant in recent times, in tandem with the scientific and technological progression of the human species. Against this backdrop, Gaian thinkers and environmentalists often get themselves into a bit of a pickle when it comes to the nature of the Universe. They often make what we can call 'the organismic fallacy':

> Human/technological domination of the planet is not compatible with the Universe being an organism that is pervaded with intrinsic value – a feeling/striving/purposeful whole.

Let us try and appreciate why environmentalists and Gaian thinkers might believe the above statement. Such an endeavour requires an exploration of the relationship between technology, environmentalism and the nature of the Universe. The first thing to explore is the issue of why it is common for Gaian thinkers and environmentalists to have an anti-technology worldview. What does it mean to have such a worldview? The anti-technology worldview is a broad view concerning what one believes to be the appropriate relationship between the human species and the Earth. Such a view entails that in an ideal world the human species would have a minimal impact on the Earth, and 'leave things to nature' as much as possible, because 'nature knows best'. In other words, technology is seen as a disruptor of pristine nature. The second thing to explore is how this anti-technology worldview can be grounded in the view that the Universe is a feeling/striving/purposeful organism. For, it is this grounding which gives rise to 'the organismic fallacy'.

It is very easy to understand why environmentalists commonly have the anti-technology worldview. Environmentalists, quite rightly, see that the range of environmental issues that currently exist are primarily caused by, or are greatly exacerbated by, technology. From our perspective, we can see that the anti-technology worldview is grounded in only one aspect of technology, its negative aspect – the temporary period of suffering and disruption which occur as a planet goes through its technological birthing process. It is crucial to come to appreciate that this negative aspect is dwarfed in significance by the positive aspect of technology. The technological birthing process produces wondrous technological fruits; the primary fruit being a new force for global cooling. The benefits and value of these fruits obviously far outweighs the disruption and suffering that occurs as these fruits are brought into being. These fruits are needed to save life on Earth. Anyone who has the anti-technology worldview obviously has not come to appreciate that technology has a cosmic purpose, it being required to save life on Earth.

People who have the anti-technology worldview will often focus on the past. They will think that life on Earth was fine before the human species evolved, fine for hundreds of millions of years, and from this they draw the erroneous conclusion that, in the absence of the human species, the future would be the same as the past. These people see no special place for the human species / technology on the Earth. Gaian thinkers will typically see the human species, and its technology, as an unwelcome cause of disruption to the planetary self-regulatory systems that have kept the Earth a habitable place for life ever since it arose on the planet. I once had the anti-technology worldview myself, in my mid-twenties, at the time before I plunged into the depths of our unfolding Solar System, and encountered its past, its present, and its future.

How does the anti-technology worldview relate to the two broad ways in which one can view the Universe? There is a popular strand in environmentalism and Gaian thinking which contends that human technological domination of the planet – domination which entails a plethora of environmental issues, immense disruption, and which has brought forth the phenomenon of human-induced global warming – has arisen because of the widespread existence of the view of the Universe as a clockwork mechanism. The idea is that a clockwork mechanism can be used and abused, whereas one cannot use and abuse what one sees as a feeling/striving/purposeful entity. One cannot justly abuse one's brothers and sisters, so the domination of the planet by the human species requires it to see its fellow planetary life-forms, from cows to trees, not as brothers and sisters, but as alien entities that are of a totally different kind to itself. All of this seems to be obviously true. It surely is true. Of course it is true! This line of thought has led some environmentalists and Gaian thinkers to call for an urgent change to the way that we view the Universe; they beseech us to once again see the Universe as an intrinsically valuable organism – a feeling/striving/purposeful whole; for, they believe that if such a worldview is widespread, that we would inevitably stop dominating the planet and environmental issues would thereby resolve themselves.

So, there is a very obvious link between the anti-technology worldview, and the view of the Universe as a clockwork mechanism – the latter is a necessary requirement for the bringing forth of technology. This means that if one has the anti-technology worldview, one will naturally be disposed to be opposed to the underpinning clockwork mechanism view which enables technology to be brought into being and to dominate its planet. It is a short-step from here to the 'organismic fallacy'. For, the view that the Universe is an organism that is pervaded with intrinsic value – a feeling, striving, and purposeful whole, is taken to be the antidote to the prescribed problem. Surely, it is believed, if the human species were to now readopt this view – the view of our ancient ancestors – it would then respect the Earth and all of its inhabitants; the human species would act very differently, and would not dominate the planet with technology. In this way, the view that the Universe is an organism that is pervaded with intrinsic value – a feeling, striving, and purposeful whole, is envisioned as being antithetical to human/technological domination of the planet. Technology is conceptualised as disrupting the purposeful intrinsically valuable unfolding organism that is the Universe, rather than being a phenomenon that contributes to its wholeness. This is 'the organismic fallacy'.

The 'organismic fallacy' is a fallacy because appreciating that the Universe is pervaded with intrinsic value, that it is a feeling/striving/purposeful whole, can just as easily lead to the opposite conclusion. It can lead to the conclusion that technology, and the human domination of the planet that is required to bring forth technology, are extremely valuable and precious parts of the unfolding purposeful whole. Indeed, this alternative conclusion is at the very heart of mystical environmental philosophy.

The anti-technology strand in environmentalism and Gaian thinking is, of course, right in its assertion that the clockwork mechanism view enables the bringing forth and complexification of technology. In other words, the clockwork mechanism view is responsible for creating the 'environmental crisis' and human-induced global warming. Where this strand of thinking goes awry is the assumption that environmental disruption is a sign that things have gone wrong at the planetary level. In a similar vein, this strand of thinking goes awry by assuming that the epoch of delusion, in which the human species temporarily comes to see the Universe as a clockwork mechanism, is a bad thing. In reality, the widespread temporary deluded belief that the Universe is a clockwork mechanism, and the associated bringing forth of technology, are signs that the Earth is absolutely thriving; they are not signs that things have gone wrong on the Earth. The Earth is simply going through a period of disruption as a precursor to moving into a more optimal state, a state in which a new force for global cooling, a technological force for global cooling, enables a long and glorious future for life on Earth.

Our temporary delusion has a purpose. Technology has a purpose. Environmental disruption has a purpose. Environmentalism has a purpose. All of these things are essential parts of the intrinsically valuable, purposeful, unfolding whole that is the Universe. The reality is that we are living though the epoch in which life on Earth brings forth technology, via the activities of the human species, in order to give itself a long and glorious future. This epoch is marked by delusion; the human species has to see its surroundings as very different to itself, as mere matter, as resources which are there for its use. This conception, the seeing of the rest of the Universe as very different to itself, is a delusion because in reality the human species is immensely similar to all of the other life-forms that exist on the Earth, and is also similar to everything else in the Universe. The human species has had to be deluded, because in the absence of this delusion the human species would not be able to bring itself to make the massive transformations to the Earth, including the mass utilisation of its fellow life-forms, which are a necessary aspect of life on Earth bringing forth its technological force for global cooling.

Have you noticed an oddity that exists within the view that the Universe is a clockwork mechanism that is devoid of intrinsic value, feeling and purpose? The clockwork mechanism view is a view of the *entire* Universe. And, humans are part of the Universe. So, this view barely makes any sense; indeed, it seems to be inherently self-contradictory. For, humans know both that they are intrinsically valuable, and that they are feeling/striving/purposeful entities. Humans also know that they are part of the Universe. So, how can humans end up viewing the Universe as being devoid of that which they know for certain exists?! This bizarre aberration, this powerful delusion that comes into existence as a planet goes through its technological birthing process, is needed. The human species has to come to see everything in the Universe that is not itself, as radically different to itself. Yet, deep within every human is to be found the knowledge that such a seeing is profoundly spurious. For, deep within every human is to be found an immense plethora of connections between the striving feelings that lie within and the striving feelings that exist without.

The non-human life-forms of the Earth have played a pivotal role in enabling life on Earth to bring forth technology. Firstly, they brought the human species into existence. Secondly, they have supported the human species in a plethora of ways. They have been trusted friends and companions. They have been sources of joy and wonder. They have been food sources that have provided an invaluable source of nutrition which has enabled cultural unfoldment to power forward towards the bringing forth and complexification of technology. They have been pivotal to the powering forward of scientific knowledge, through being subjects of dissection and experimentation. You might find this to be extremely distasteful, and like me you might be a long-term vegan; however, one can also accept that all of these activities are an inevitable transitory part of our unfolding Earth / Solar System as it progresses to a better future.

This is a future in which all life-forms are respected by the human species. It is a future in which the atrocious age of the technological explosion, with all of its immense disruption, pain and suffering, is now a distant memory. It is a future in which the life that has arisen on the Earth still exists. It is a future in which the human species is no longer deluded. It is a future in which life on Earth is positively thriving because the human species has fulfilled its cosmic purpose through the deployment of solar-deflection technology at Lagrangian Point L1.

The Oceanic Travelling Carbon Dioxide

Gushes Out into a

Vastly Changed World

The Oceanic Travelling Carbon Dioxide Gushes Out into a Vastly Changed World

Ecstasy!   What have you done?

You have released the fruits of the first line of defence!!

Some of this carbon dioxide is currently in the Earth's atmosphere. It has enhanced the 'greenhouse effect' and thereby caused the Earth's atmospheric temperature to rise.

The rising atmospheric temperature has caused the oceanic temperature to rise.

The rising atmospheric temperature has caused the melting of icebergs and glaciers in the North Atlantic and Greenland.

I wonder: What has happened to the rest of the fruits that you released from safe storage?

These fruits, a ginormous amount of carbon dioxide, got immediately sucked into the oceanic thermohaline circulation. They have been there ever since, biding their time, waiting to be unleashed into the Earth's atmosphere.

Over the next few hundred years these fruits will end their oceanic adventure; they have been on vacation since the start of the Industrial Revolution; they are about to return to the Earth's atmosphere, the place from which they departed so very long ago.

Yet, it is a vastly different world to which they are returning.

Where have the rainforests gone!

Megacities, cars, airplanes – What on earth are these strange things?!!

Why are the oceans so much warmer now? The carbon dioxide was easily able to absorb itself into the cooler water that existed in the world that it departed from. Now that the oceans are warmer it will have to stay in the atmosphere!!

Why is the thermohaline circulation so weak now? Have the icebergs and glaciers melted? The thermohaline circulation is so slow that the returning carbon dioxide is unable to be absorbed back into the ocean. It will have to stay in the atmosphere!!

The oceanic travelling carbon dioxide is about to make a dramatic return into a vastly changed world. As it starts to gush out into the Earth's atmosphere over the next few hundred years, the carbon dioxide concentrations in the atmosphere are set to spiral upwards. There is nowhere else for it to go in a world of transformed land surfaces and transformed oceans.

The Earth's atmospheric temperature is set to shoot up. Positive feedback loops are set to kick in. The Earth is on course to become an inhospitable and barren place.

Wait: Is there any hope for life on Earth? Has Hölderlin's "Mysterious Prince" arrived?!

Will ecstasy be successful?

Is the precious saviour of life on Earth fulfilling its cosmic purpose?

If life on Earth takes technological control of the Earth's atmospheric temperature, then it can keep the atmospheric temperature down, keep it favourable for life, when the oceanic travelling carbon dioxide makes its imminent and dramatic return.

But if the human species maintains its attempt to make a sheepish retreat, if it wastes time faffing around attempting to cut fossil fuel emissions, if it shuns the technologies that are deployed and sufficiently deployed, then the precious life that has arisen on the Earth will be obliterated.

9 781907 962882